IODP-Canada Summer School on Ocean and Climate Changes in Polar and Subpolar Environments 2010

IOP Conference Series: Earth and Environmental Science Volume 14

Montreal, Quebec, Canada
27 June – 12 July 2010

ISBN: 978-1-5108-0210-0
ISSN: 1755-1307

Printed from e-media with permission by:

Curran Associates, Inc.
57 Morehouse Lane
Red Hook, NY 12571

Some format issues inherent in the e-media version may also appear in this print version.

Printed by Curran Associates, Inc. (2015)

For permission requests, please contact the Institute of Physics
at the address below.

Institute of Physics
Dirac House, Temple Back
Bristol BS1 6BE UK

Phone: 44 1 17 929 7481
Fax: 44 1 17 920 0979

techtracking@iop.org

Additional copies of this publication are available from:

Curran Associates, Inc.
57 Morehouse Lane
Red Hook, NY 12571 USA
Phone: 845-758-0400
Fax: 845-758-2634
Email: curran@proceedings.com
Web: www.proceedings.com

TABLE OF CONTENTS

The great challenges in Arctic Ocean paleoceanography

Ruediger Stein

Alfred Wegener Institute for Polar and Marine Research, 27568 Bremerhaven, Germany

E-mail: Ruediger.Stein@awi.de

Abstract. Despite the importance of the Arctic in the climate system, the data base we have from this area is still very weak, and large parts of the climate history have not been recovered at all in sedimentary sections. In order to fill this gap in knowledge, international, multidisciplinary expeditions and projects for scientific drilling/coring in the Arctic Ocean are needed. Key areas and approaches for drilling and recovering undisturbed and complete sedimentary sequences are depth transects across the major ocean ridge systems, i.e., the Lomonosov Ridge, the Alpha-Mendeleev Ridge, and the Chukchi Plateau/Northwind Ridge, the Beaufort, Kara and Laptev sea continental margins, as well as the major Arctic gateways towards the Atlantic and Pacific oceans. The new detailed climate records from the Arctic Ocean spanning time intervals from the Late Cretaceous/Paleogene Greenhouse world to the Neogene-Quaternary Icehouse world and representing short- and long-term climate variability on scales from 10 to 10^6 years, will give new insights into our understanding of the Arctic Ocean within the global climate system and provide an opportunity to test the performance of climate models used to predict future climate change. With this, studying the Arctic Ocean is certainly one of the major challenges in climate research for the coming decades.

1. Introduction and background

Although major progress in Arctic Ocean research has been made during the last decades, the knowledge of its short- and long-term paleoceanographic and paleoclimatic history as well as its plate-tectonic evolution is much behind that from the other world's oceans. That means - despite the importance of the Arctic in the climate system (see [1] for recent review and references) - the data base we have from this area is still very weak, and large parts of the climate history have not been recovered at all in sedimentary sections (figure 1). This lack of knowledge is mainly caused by the major technological/logistic problems in reaching this permanently ice-covered region with normal research vessels and in retrieving long and undisturbed sediment cores.

Concerning future Arctic geoscientific research, the climate history must be studied at different temporal scales from 10 to 10^6 years, as paleoclimate research and climate models demonstrate that processes and varying conditions in polar regions play a key role in driving and amplifying global climate variability and sea-level change on time scales of decades to millions of years. For example, the Arctic sea ice, a key indicator and agent of climate change, affecting surface reflectivity, cloudiness, humidity, exchanges of heat and moisture at the ocean surface, and ocean currents, had been undergoing retreat over the past three decades, as recognized by the science community with some alarm [2]. In order to contextualize recent dramatic reductions in sea ice cover in the polar regions, it is essential to determine spatial and temporal changes in sea ice occurrence and its natural variability in the past [3, 4]. To reconstruct past climate (including sea ice) variability and the interrelation of different processes for times prior to direct measurements, one has to rely on indirect evidence, i.e., on information provided by "proxies" determined in sediment (and ice) cores. These proxies may be used for reconstruction of past temperature, salinity, sea ice cover, global ice volume, nutrients, marine biological productivity, etc. [1, 5]. Whereas a large number of these proxies are routinely used, others are new and still have to be fully established in Arctic Ocean environmental research. In this context, one key issue of overall importance in Arctic research is the establishment of a stratigraphic/chronological framework at highest accuracy possible, to be achieved with a combined effort of different stratigraphic methods including radiometric dating, microfossils, stable isotopes, cosmogenic isotopes, and geomagnetic records. Such a framework is the prerequisite for all paleoenvironmental studies and for the correlation of Arctic sedimentary records with other sedimentary records from the low, mid, and southern high latitudes, as well as climate records from land and ice cores ([1] and references therein).

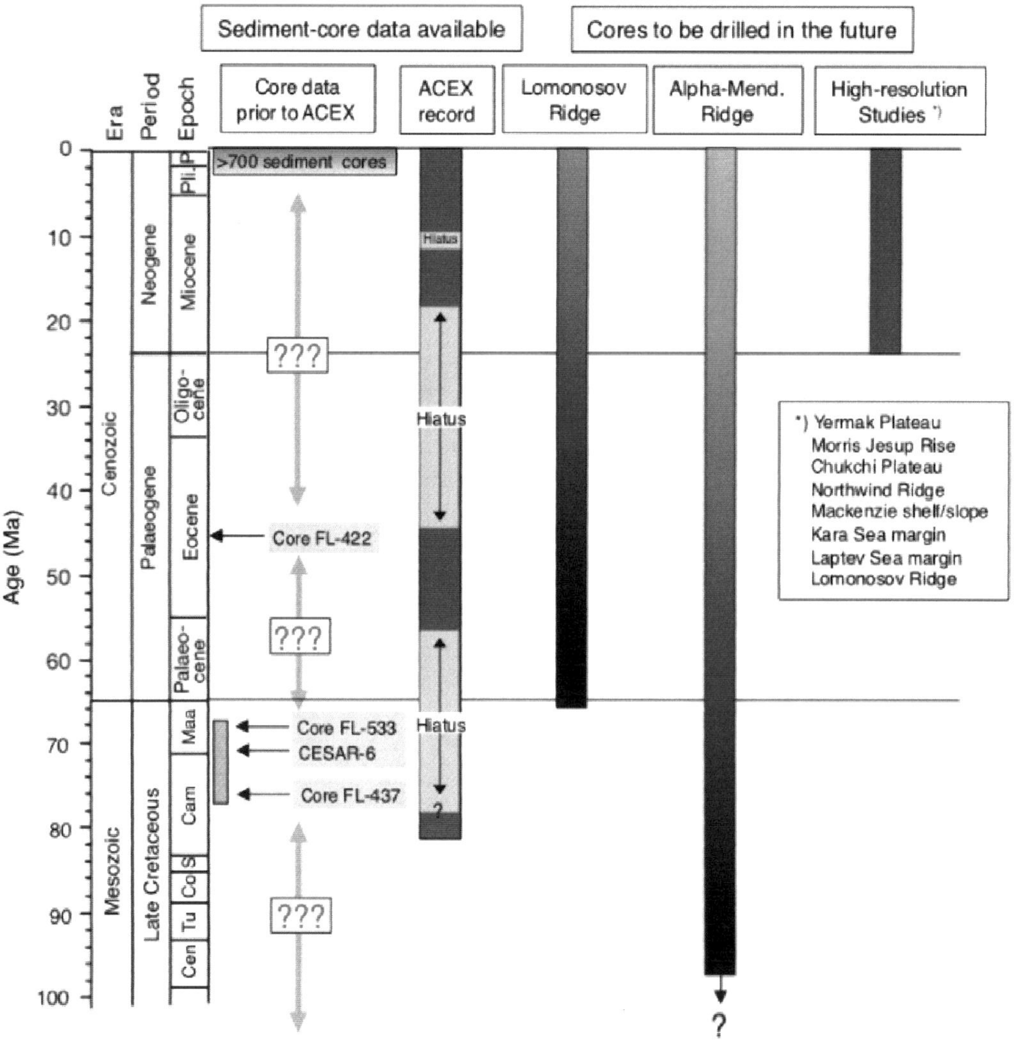

Figure 1. Stratigraphic coverage of existing cores in the central Arctic Ocean prior to ACEX (based on [6]) and the section recovered during the ACEX drilling expedition [7, 8], and stratigraphic coverage and key locations of sites to be drilled in the future [1].

2. Future major research themes in Arctic Ocean paleoceanography

Some of the key topics of future research are presented here. These topics are certainly not complete and have to be seen as examples.

2.1. Quaternary and Neogene climate variability on sub-millenial to Milankovich time scales

Arctic high-resolution (sub-millenial) climate records are still rare and known from only a few locations in the Nordic Seas and the Eurasian shelf seas. Some of these high-resolution records, for example, indicate short-term centennial- to millennial-scale variability in Arctic climate and river discharge correlating with the Greenland ice core record, changes in sea-ice cover in the North Atlantic and advance/retreat of Norwegian glaciers, and may reflect natural cyclic climate variations related to changes of the North Atlantic Oscillation/Arctic Oscillation (NAO/AO) pattern as described for the last decades [1, 9]. Considering the importance of the Arctic Ocean for the global ocean circulation and the climate system, more high-resolution paleoceanographic data sets from key areas are needed to reconstruct the climatically important parameters (e.g., sea-surface temperature and salinity, inflow of Atlantic Water, sea-ice cover, river discharge). This kind of data sets going back beyond the time scale of direct discharge measurements, can be used to determine the natural variability of these parameters as a background for an assessment of anthropogenically influenced changes in the last 100-150 years.

In this context, the strategic approach is to merge atmosphere and ocean signals from ice, marine and land (lake/permafrost) climate records at orbital, millennial, and up to decadal time scales to decipher their complex inter-relationship and impact, their relation with processes in low and middle latitudes, and the effect of external forcing. Of overall interest is the understanding of the polar mechanisms and thresholds in triggering rapid (10^2-10^3 yr) climate changes during warm and cold climate conditions (ice-permafrost-ocean-atmosphere mechanisms) and the comparison of the spatial and temporal evolution of such changes in Northern and Southern High Latitudes.

High-resolution records are needed from the Fram Strait/Yermak Plateau area (sites to study the variability of North Atlantic water inflow, sea-ice cover, primary production, etc.), the continental margins surrounding the Arctic Ocean (sites to study the variability of circum-Arctic river discharge, marginal-ice-zone processes, Atlantic- and Pacific-water influence, etc.), and the central Arctic Ocean (sites to study the variability of sea-ice cover and drift, water-mass characteristics, primary production, etc.). Coring sites characterized by high sedimentation rates, have to be carefully selected by bathymetric (Hydrosweep) and acoustic/seismic (PARASOUND and multi-channel seismic) surveys. For studying the Holocene climate variability, cores may be retrieved by conventional piston and kastenlot coring. For studying the (sub-) millennial climate variability on longer time scales, i.e., under different boundary conditions during Quaternary/Neogene times, however, drill cores are needed.

2.2. Long-term Mesozoic-Cenozoic climate history of the Arctic Ocean

Despite the success of Integrated Ocean Drilling Program (IODP) Expedition 302 -ACEX- [7, 10], major key questions related to the climate history of the Arctic Ocean and its long- and short-term variability during Mesozoic-Cenozoic times, cannot be answered from the ACEX record due to the poor core recovery and, especially, a major mid-Cenozoic hiatus (figure 1). This hiatus just spans the critical time when prominent changes in global climate took place during the transition from the early Cenozoic Greenhouse world to the late Cenozoic Icehouse world (figure 2; [11] and references therein). The success of ACEX has certainly opened the door for further scientific drilling in the Arctic Ocean. The ACEX results will frame the next round of questions to be answered from new drill holes to be taken during a series of drilling legs. Some of the main scientific topics are listed in the following (from [1] and references therein):

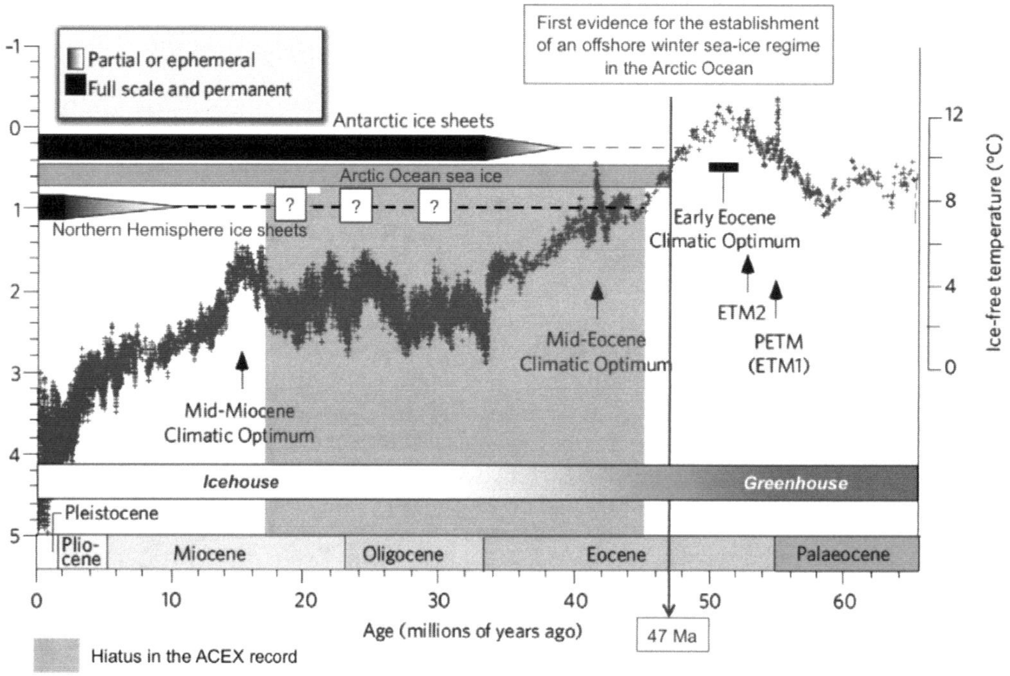

Figure 2. A smoothed global benthic foraminifer $\delta^{18}O$ time series showing the long-term cooling and the Greenhouse/Icehouse transition through Cenozoic times ([11], supplemented). The occurrence of Cenozoic glaciations on the Northern and Southern Hemisphere and Arctic sea ice are shown [11, 12]. The hiatus in the ACEX record is indicated.

(1) The Cenozoic transition of the Earth's climate from one extreme (Paleogene greenhouse lacking ice) to another (Neogene icehouse with bipolar glaciation characterized by an Antarctic continental ice-cap and seasonally variable but persistent sea-ice cover in the Arctic) is linked to increased latitudinal gradients and oceanographic changes that connected surface and deep-sea circulation between high and low latitude oceans. The general Cenozoic cooling trend, however, is interrupted by warming intervals as well as short-term extreme cooling transients [11]. Some of the related key questions are: Did the Arctic Ocean climate follow the global trend shown in figure 2? Are the Early Eocene Climate Optimum (poor recovery in the ACEX record) and the Oligocene and Mid-Miocene warmings also reflected in Arctic Ocean records? Were extensive glaciations developed synchronously in both the Northern and Southern Hemispheres? What was the variability of sea ice in terms of frequency, extent and magnitude?

(2) Black biosiliceous silty clays and clayey silts rich in organic carbon were found throughout the upper early to middle Eocene of the ACEX record [7, 10, 13], but also in a short sediment core from the Alpha Ridge, representing the late Cretaceous [14]. These data indicate a paleoenvironment characterized by poorly ventilated bottom waters and high but variable primary production. When and how did these extreme conditions develop during Mesozoic and Cenozoic times? When and how did the change to oxygenated bottom waters typical for the Neogene and Quaternary Arctic Ocean occur? Did anoxia occur down to the deep basins?

(3) Drill sites more proximal to the Siberian margin allows a detailed study of the history of river discharge and its paleoenvironmental significance. In this context, the Miocene uplift of the

Himalayan-Tibetan region is of particular interest as it may have triggered enhanced flow of Siberian rivers and changed the fresh-water balance of the Arctic's surface waters, considered to be a key factor for the formation of Arctic sea-ice and onset of major glaciations [15], a hypothesis to be tested by drilling along the Kara and Laptev Sea continental margin influenced by discharge of the large rivers Lena, Ob, and Yenisei.

(4) How critical is the exchange of water masses between the Arctic Ocean and the Atlantic and Pacific oceans for the long-term climate evolution as well as rapid climate change?

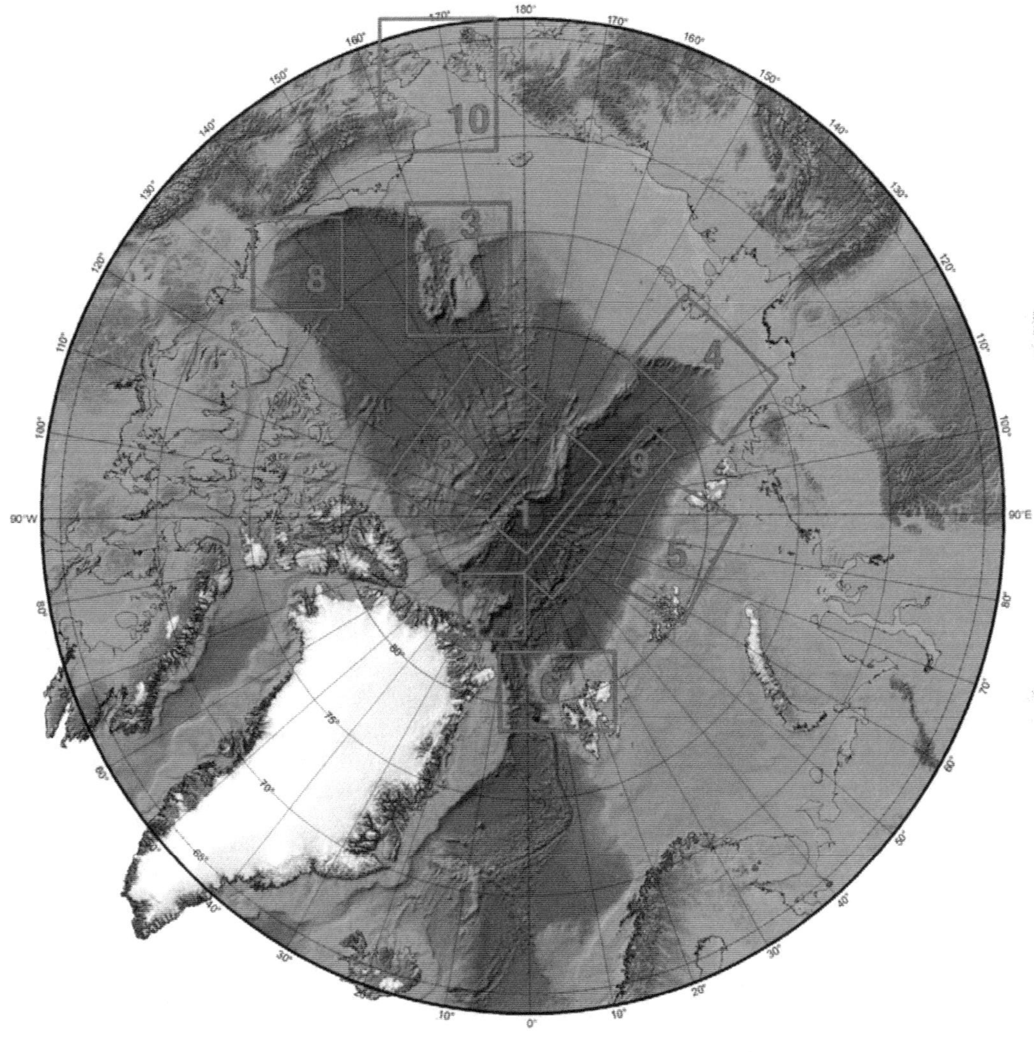

Figure 3. Key areas for future drilling areas in the Arctic Ocean (based on [16, 17]; bathymetric map according to [18]). 1 = Lomonosov Ridge; 2 = Alpha-Mendeleev Ridge; 3 = Chukchi Plateau/Northwind Ridge; 4 = Laptev Sea continental margin; 5= Kara Sea continental margin; 6 = Fram Strait/Yermak Plateau; 7 = Morris Jesup Rise; 8 = Mackenzie shelf/slope; 9 = Gakkel Ridge; 10 = Northern Bering Sea/Bering Strait area.

3. Key areas for future Arctic Ocean drilling

In order to study the long-term Mesozoic-Cenozoic climate evolution, we need to obtain undisturbed and complete sedimentary sequences to be drilled on depth transects across the major ocean ridge systems, i.e., the Lomonosov Ridge, the Alpha-Mendeleev Ridge, and the Chukchi Plateau/Northwind Ridge (figure 3, key areas 1 to 3). High-resolution records allowing to study climate variability on Milankovich and millennial to sub-millennial time scales, can be drilled along the continental margins characterized by high sedimentation rates. Here, key areas are the Kara, Laptev and Beaufort seas characterized by their large river discharge (figure 3, key areas 4, 5 and 8). Key location for studying the history of exchange of the Arctic Ocean with the world's oceans are the Fram Strait/Yermak Plateau, Morris Jesup Rise and Chukchi Plateau/Northwind Ridge and Bering Sea areas (figure 3, areas 3, 6, 7, and 10).

4. Concluding remarks and outlook

Concerning the short- and long-term evolution of the Arctic Ocean and its importance for the understanding of the global climate history, most of the key questions mentioned above as well as the key areas for scientific drillling in the Arctic Ocean needed to answer these questions, were already identified on several workshops during the last two decades and published in workshop reports [16, 17, 19, 20, 21]. Over the years, however, scientific drilling in the ice-covered Arctic Ocean remained a dream. The ACEX drilling in 2004 [7, 10] was the first major step that part of this dream became reality. Now, further drilling campaigns including the use of new technologies, are needed to follow-up in the future. In this context, for example, the "MeBo", a remotely controlled drill rig that is deployed on the sea bed, and developed at the Center for Marine Environmental Sciences (MARUM), Bremen University, Germany, might be an option for future Arctic drilling (for technical details see http://www.marum.de/en/Sea_floor_drill_rig_MeBo.html). MeBo has been successfully used in a recent drilling campaign at the continental slope off Chile where more than 70 m long continuous sediment records with an average core recovery of close to 90% were obtained, and it is planned to further develop MeBo to be suitable even for drilling longer cores up to 200 m [22]. MeBo-type drilling could be carried out from existing ice-strengthend research vessels (such as the German *Polarstern,* a double-hulled icebreaker that can break through ice 1.5 metres thick at a speed of approx. 5 knots). That means, this type of Arctic drilling which is certainly sufficient for reaching numerous major goals in Arctic Ocean paleoceanographic research, could even start almost now and allow an IODP-type drilling program. In order to obtain long (>500 m) drill cores from the oceanic ridges as well as the deep basins, however, the construction of a new large icebreaker with deep-water drilling capability - such as the *Aurora borealis* – might be the optimum, but also very expensive solution that may open another new dimension in multidisciplinary Arctic Ocean research [23].

Finally, it should be clearly emphasized that for the precise planning of any future Arctic Ocean drilling campaigns including site selection, evaluation of proposed drill sites for safety and environmental protection aspects, etc., comprehensive site survey data are needed first. That means, the development of detailed site survey strategy is a major challenge for the coming years.

References

[1] Stein R 2008 *Arctic Ocean Sediments: Processes, Proxies, and Palaeoenvironment (Developments in Marine Geology* vol 2*)* (Amsterdam: Elsevier) pp 1–587

[2] Serreze M C, Holland M M and Stroeve J 2007 Perspectives on the Arctic's shrinking sea-ice cover *Science* **315** 1533–36

[3] Massé G, Rowland S J, Sicre M A, Jacob J, Jansen E and Belt S T 2008 Abrupt climate changes for Iceland during the last millennium: Evidence from high resolution sea ice reconstructions *Earth Planet. Sci. Lett.* **269** 565–69

[4] Mueller J, Massé G, Stein R and Belt S 2009 Extreme variations in sea ice cover for Fram Strait during the past 30 ka *Nature Geoscience* doi: 10.1038/NGEO665c

[5] Hillaire-Marcel C and de Vernal A (ed) 2007 *Proxies in late Cenozoic paleoceanography (Developments in Marine Geology* vol 1*)* (Amsterdam: Elsevier) pp 1–843
[6] Thiede J, Clark D L and Hermann Y 1990 Late Mesozoic and Cenozoic paleoceanography of the northern polar oceans *The Arctic Ocean Region (The Geology of North America* vol L*)* ed A Grantz *et al* (Boulder: Geological Society of America) pp 427–58
[7] Backman J, Moran K, McInroy D B, Mayer L A and the Expedition 302 Scientists 2006 *Proceedings IODP* vol 302 (College Station: Integrated Ocean Drilling Program Management International) doi:10.2204/iodp.proc.302.104.2006
[8] Backman J *et al.* 2008 Age Model and Core-Seismic Integration for the Cenozoic ACEX Sediments from the Lomonosov Ridge *Paleoceanography* doi:10.1029/2007PA001476
[9] Dickson R R, Osborn, T J, Hurrell J W, Meincke J, Blindheim J, Adlandsvik B, Vinje T, Alekseev G and Maslowski W 2000 The Arctic Ocean response to the North Atlantic Oscillation *J. Clim.* **13** 2671–96
[10] Moran K, Backman J *et al* 2006 The Cenozoic palaeoenvironment of the Arctic Ocean *Nature* **441** 601–5
[11] Zachos J C, Dickens G R and Zeebe R E 2008 An early Cenozoic perspective on greenhouse warming and carbon-cycle dynamics *Nature* **451** 281–83
[12] Stickley C E, St John K, Koc N, Jordan R W, Passchier S, Pearvce R B and Keasrns L E 2009 Evidence for middle Eocene Arctic sea ice from diatoms and ice-rafted debris *Nature* **460** 376–80
[13] Stein R, Boucsein B and Meyer H 2006 Anoxia and high primary production in the Paleogene central Arctic Ocean: First detailed records from Lomonosov Ridge *Geophys. Res. Lett.* **33** L18606 doi:10.1029/2006GL026776
[14] Clark D L, Byers C W and Pratt L M 1986 Cretaceous black mud from the central Arctic Ocean *Paleoceanography* **1** 265–71
[15] Driscoll N W and Haug G H 1998 A Short Circuit in Thermohaline Circulation: A Cause for Northern Hemisphere Glaciation? *Science* **282** 436–38
[16] Stein R and Coakley B 2009 Scientific Drilling in the Arctic Ocean: A challenge for the next decades *INVEST White Paper* http://www.marum.de/Binaries/Binary42266/SteinArcticOcean.pdf
[17] Coakley B and Stein R 2010 Arctic Ocean Scientific Drilling: The Next Frontier *Scientific Drilling* **9** 45–9
[18] Jakobsson M, Macnab R, Mayer L, Anderson R, Edwards M, Hatzky J, Schenke H W and Johnson P 2008 An improved bathymetric portrayal of the Arctic Ocean: Implications for ocean modeling and geological, geophysical and oceanographic analyses *Geophys. Res. Lett.* **35** L07602, doi:10.1029/2008GL033520
[19] Thiede J and the NAD Science Committee 1992 The Arctic Ocean Record: Key to Global Change (Initial Science Plan) *Polarforschung* **61** 1–102
[20] Hovland M (ed) 2001 The High-Arctic Drilling Challenge *Final Report of the Arctic's Role in Global Change Program Planning Group (APPG)* (College Station: Ocean Drilling Program) pp 1–38
[21] Bowden S, Corell R W, Hassol S J and Symon, C (ed) 2007 *Arctic Research: A global responsibility: Report of ICARP-II* (Montreal: McCallum Printing Group Inc) pp 1-36
[22] Freudenthal T, Hebbeln D, Mohtadi M, Martinez-Méndez G and Wefer G 2011 70 m double hole drillings for paleoceanographic research with the sea floor drill rig MeBo *Geophys. Res. Abstr.* **13** EGU2011-3650-1
[23] Thiede J and Egerton P 2004 *AURORA BOREALIS: A Long-Term European Science Perspective for Deep Arctic Ocean Research 2006-2016* (Strasbourg: European Science Foundation, N° d'ordre 042478) pp 1–80

IOP Conf. Series: Earth and Environmental Science **14** (2011) 012002 doi:10.1088/1755-1315/14/1/012002

Late Cenozoic Paleoceanography of the Central Arctic Ocean

Matt O'Regan [1,2]

[1]School of Earth and Ocean Sciences, Cardiff University, Main Building, Park Place, Cardiff, CF10 3AT, Wales, UK

[2]Department of Geological Sciences, Stockholm University, SE-106 91, Stockholm Sweden

E-mail: oreganM1@cardiff.ac.uk

Abstract The Arctic Ocean is the smallest and perhaps least accessible of the worlds oceans. It occupies only 26% of the global ocean area, and less than 10% of its volume [1]. However, it exerts a disproportionately large influence on the global climate system through a complex set of positive and negative feedback mechanisms directly or indirectly related to terrestrial ice and snow cover and sea ice. Increasingly, the northern high latitude cryosphere is seen as an exceptionally fragile part of the global climate system, a fact exemplified by observed reductions in sea ice extent during the past decades [2]. The paleoceanographic evolution of the Arctic Ocean can provide important insights into the physical forcing mechanisms that affect the form, intensity and permanence of ice in the high Arctic, and its sensitivity to these mechanisms in vastly different climate states of the past. However, marine records capturing the late Cenozoic paleoceanography of the Arctic are limited - most notably because only a single deep borehole exists from the central parts of this Ocean. This paper reviews the principal late Cenozoic (Neogene/Quaternary) results from the Arctic Coring Expedition to the Lomonosov Ridge and in light of recent data and observations on modern sea ice, outlines emerging questions related to three main themes: 1) the establishment of the '*modern*' Arctic Ocean and the opening of the Fram Strait 2) the inception of perennial sea ice 3) The Quaternary intensification of Northern Hemisphere glaciations.

1. Introduction

The projected disappearance of perennial sea ice in the Arctic Ocean is one of the most alarming and immediately realized impacts of continued global warming [3]. Considerable uncertainty remains in the anticipated timing for this transition under future CO_2 emission scenarios [4]. Multi-model ensemble mean estimates under 'business as usual' scenarios in the IPCC 4th Assessment Report, indicate that perennial sea ice may disappear anywhere from 2050 to well beyond 2100, corresponding to atmospheric CO_2 levels of 520 to >700 ppmv [4]. However, the rate of sea ice reduction derived from satellite observations during the past decade is not captured in these models, suggesting that enhanced sensitivity of sea ice to the feedback mechanisms that amplify global temperature change in the Arctic may remain poorly understood and paramaterized [5,6].

One of the most illustrative examples of sea ice sensitivity to global climate change was presented by *Johannessen* [7], by comparing the satellite derived annual sea ice extent from 1979-2007 against atmospheric CO_2 concentrations during the same time period (Figure 1). The strong linear correlation between these datasets was used to forecast the sea ice extent up until 2050, and illustrated that the corresponding predictions are several million km^2 lower than IPCC ensemble mean predictions.

1 To whom any correspondence should be addressed

Published under licence by IOP Publishing Ltd

On geologic timescales, the sensitivity of the Arctic to climate change was recently quantified through a comparison of reconstructed Arctic and average Northern Hemisphere (NH) summer temperatures during four periods in the last 3 Million years (the Holocene thermal maximum, the last glacial maximum, the last interglacial and the mid-Pliocene). The results illustrate that Arctic temperature anomalies are amplified by a factor of 3-4 above the NH average during periods when the global climate cools and warms [6]. However, we still lack a complementary understanding of the variability of sea ice in the geologic past, or how the mechanisms that control it have changed. The striking correlation between sea ice extent and atmospheric CO_2 in Figure (1), and the apparently predictable amplification of temperature change in the Arctic, raise questions about how useful these relationships are for hind-casting when large-scale boundary conditions are altered (Figure 2)

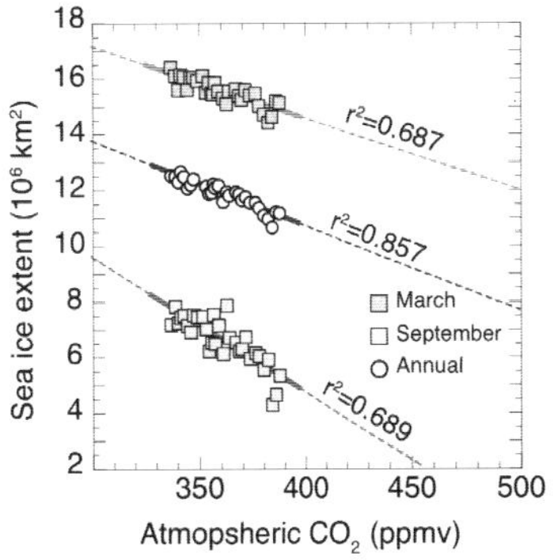

Figure 1. Arctic sea ice extent and CO_2. Annual, September and March Arctic sea ice extent for the period between 1979-2007 [7] shown against atmopsheric CO_2 concentraions from the Mauan Loa Observatory (wwwesrlnoaagov/gmd/ccgg/trends/).

In addition to elevated (yet poorly constrained) CO_2 concentrations in the Cenozoic, these boundary conditions include changes in the freshwater budget, and exchange with both the Pacific and Atlantic Oceans. These changes clearly modify atmospheric and oceanic circulation patterns and impact the vertical stratification and overall heat budget of the Arctic. Similarly, the advance and retreat of large ice sheets alter regional temperature and circulation patterns and have certainly changed dramatically through the late Cenozoic. An understanding of how these boundary conditions have influenced the form, intensity and permanence of ice cover in the Arctic can help improve our understanding of the complex modern ocean-atmosphere-ice system and how it has evolved with global climate. It is also needed to define natural variability within this system and identify periods in the geologic past that may serve as potential analogues for future climate states. As it is the oceanic response that is of greatest concern, these insights must come from the analysis of marine sediments

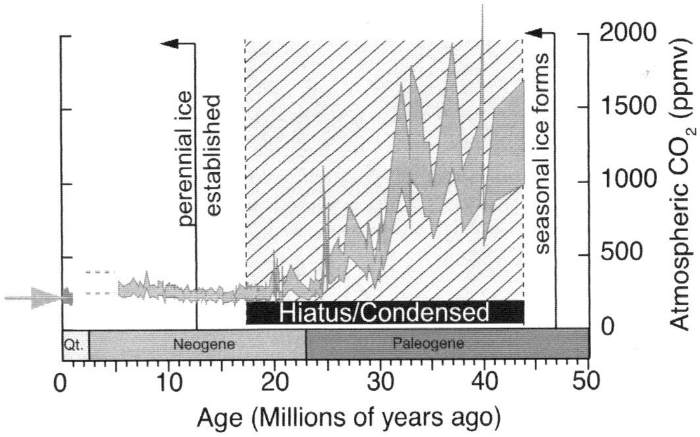

Figure 2. Cenozoic reconstruction of atmospheric carbon dioxide concentration from marine proxy records [9]. Grey arrow on left indicates pre-industrial CO_2.. Results from the Arctic Coring Expedition on the Lomonosov Ridge indicate that seasonal sea ice was established around 47 Ma, when greenhouse gas concentrations were comparable to levels at 2-4x the modern. ACES results also suggest the formation of perennial sea ice in the Miocene, coincident with or prior to the Middle Miocene Climate Optimum (See section 4). (. Quaternary (Qt). **It is important to note that ongoing debates exist about the actual magnitude and variability of CO_2 through the Cenozoic*

2. Coring and drilling in the Arctic

It was only in 1991 that the first non-nuclear powered icebreakers (the Swedish *R/V Oden* and German *R/V Polarstern*) reached the North Pole. Since this time, hundreds of cores were retrieved from the central Arctic Ocean and its Marginal seas, adding considerably to the inventory of cores collected in previous decades, many of which from floating ice camps (Figure 3). Difficulties in dating these sediments, and the short nature of the recovered sequences (generally less than 10-15 m) has limited our ability to characterize paleoceanographic changes occurring before the last 200-300 thousand years and address the long-term stability of sea ice under different orbital, climatic and tectonic boundary conditions.

In general terms, difficulties in dating arise from the often microfossil poor glaciomarine nature of the sediments, and is compounded by a complex downhole paleomagnetic signal that has yet to be integrated with the global geomagnetic timescale. These problems are generally not as severe in marginal settings of the Arctic Ocean. Similar challenges are faced when attempting to derive proxies for past sea ice conditions, and many established and emerging proxies, including microfossil based transfer functions [10], abundances [11] and organic biomarkers [12] have yet to be successfully applied in central Arctic Ocean sediments.

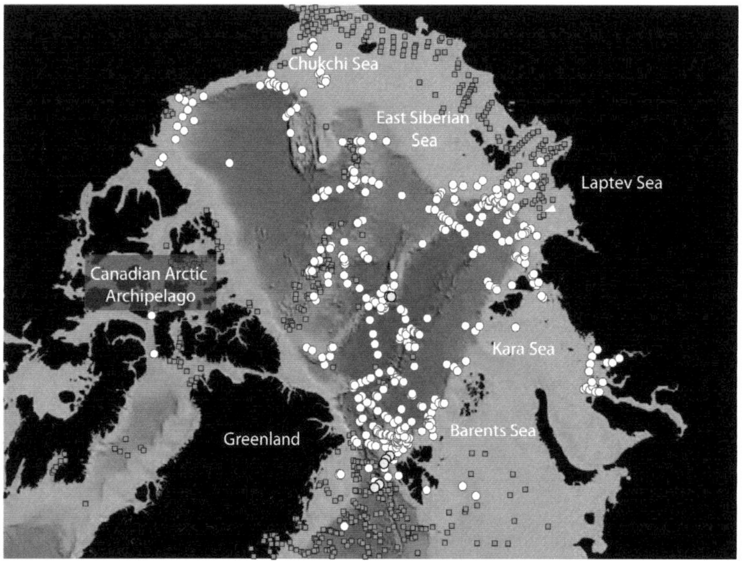

Figure 3. Partial inventory of gravity and piston cores from the Arctic Ocean. Green squares = data downloaded from wwwgeomapapporg. Red circles = compiled data from icebreaker led expeditions in the past 20 years. Yellow circles = ODP/IODP boreholes Detailed data from the analysis of many of these cores can be found at wwwpangaeade. For color reporductions of this and other figures, the reader is referred to the online version of this article.

An important realization for anyone studying the Cenozoic paleoceanography of the Arctic, is that while absolute and relative ages remain difficult to determine, correlating depositional sequences between cores and across relatively large spatial distances is possible using continuous downhole records of the lithologic, magnetic, and physical properties of the sediments and their microfaunal composition [13]. Depending on how robust the stratigraphic correlation is, this approach can be used to identify syndepositional sediments and combine age markers derived from different cores (Figure 4). Although not yet exploited, these correlations can also be used to investigate changes in the vertical water mass properties on glacial/interglacial timescales without needing a detailed or finely tuned age-model (Figure 4).

Despite the continued advance of our knowledge on late Quaternary paleoceanography in the Arctic (*see* [19] for a detailed overview), the largest contribution to our emerging understanding of late Cenozoic Arctic Paleoceaonography came from the Integrated Ocean Drilling Program's pioneering Arctic Coring Expedition (ACEX). This expedition not only recovered the first Cenozoic marine

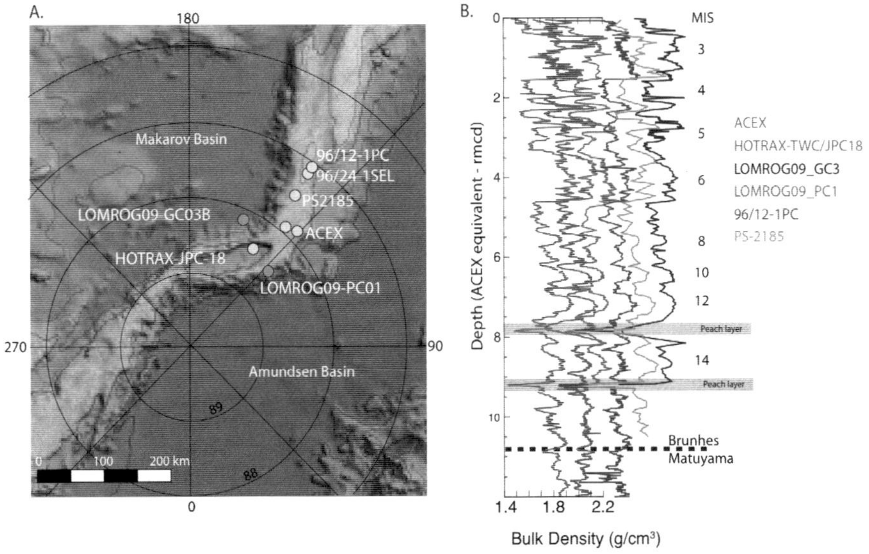

Figure 4. A. Bathymetric map of the central Lomonosov Ridge [14] showing the location of six cores collected from varying water depths (Table 1) that can be stratigraphically correlated without prior knowledge on the age of the sediments. **B.** Stratigraphic correlation of the six cores illustrated using downhole bulk density profiles. The cores are migrated onto the ACEX depth scale using prominent tie points. Inferred ages are from compiled relative and absolute dating techniques applied to the different records and covering MIS 1-6 (Table 1), with older ages derived from cyclostratigraphic analysis of the ACEX record [13]. **C.** A number of different chronostratigraphic techniques were applied to these cores to acquire ages for sediments during the last 200 Kyr. These are illsutrated and color-coded to show which core they were derived from (Table 1). While some scatter still exists, there is overall agreement between the different chronostratigraphic methods back to the end of MIS5. The detail of the stratigraphic correlation from **B** is shown on the right using grain size records for the upper 6 m of the migrated depth scale.

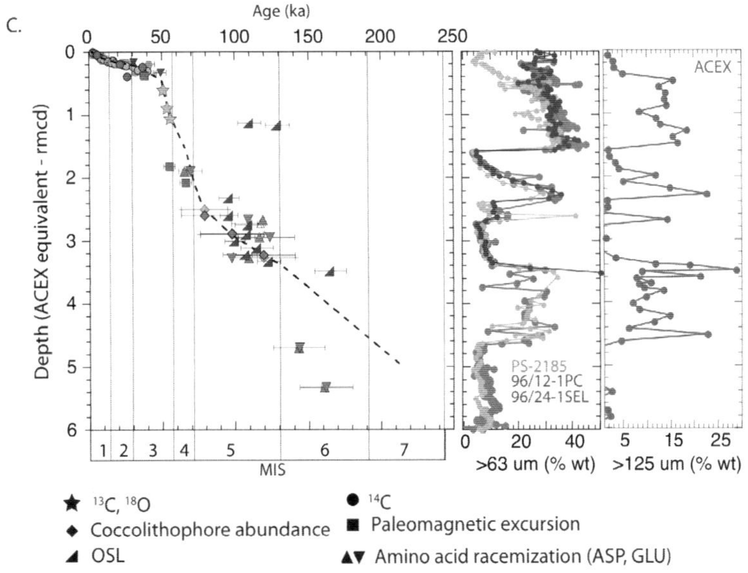

sediments sequence from the central Arctic Ocean, but illustrated that drilling was possible in the continuously moving ice pack, there-by opening the way for future scientific drilling [20]. ACEX targeted 4 closely spaced sites on the circumpolar region of the Lomonosov Ridge Collectively, a 428-meter sequence of Cenozoic sediments was penetrated and partially recovered, reflecting deposition during the subsidence and drift of the Lomonosov Ridge towards its present position [21]. Many of the traditional limitations to dating Arctic sediments, namely the sporadic and low occurrence of microfossils and a complex magnetostratigraphy, were also encountered in the ~198 m late Cenozoic section of the ACEX record. To overcome these limitations a few key biostratigraphic datums were used in conjunction with beryllium isotopes and partially supported through cyclostratigraphic analysis [22] (Figure 5).

Table 1. Locations, modern water depths, methods of dating and references for cores displayed in Figure 4 PM= Paleomagnetic excursions; cn= calcareous nannofossil abundance/assemblage data; AAR=Amino Acid Racemization; OSL=Optically Stimulated Luminescence Dating; cy=cyclostratigraphy.

Core	Lat	Lon	Water Depth (m)	Dating method	Reference
ACEX (Holes 4C/ 3A)	879	1362-1395	1200-1290	^{14}C, PM, cy	[15, 13]
96/12-1PC	871	1448	1003	AAR, cn, cy	[16]
96/24-1SEL	872	1446	980	OSL	[17]
PS-2185	875	1442	1052	^{14}C, cn, ^{13}C, ^{18}O, ^{10}Be	[13]
HOTRAX-JPC18	884	1466	2598	^{14}C, cn	[18]
LOMROG09-GC03	882	1564	3814	N/A	
LOMROG09-PC01	885	1335	1244	N/A	

One of the surprising initial findings was the presence of a 26 million year (Myr) hiatus separating early to middle Eocene (44.4 Ma) from late early Miocene sediments (18.2 Ma), a result that had not been anticipated from pre-cruise analysis of seismic data. The location and duration of this hiatus, and the overall mid Cenozoic age model, were recently challenged by osmium isotope dates (Figure 5) [23]. Findings that potentially have large implications for our understanding of when the *modern* conditions were established in the Arctic, defined by ventilated bottom-water conditions and predominantly glaciomarine sediments [21].

3. Establishment of the '*modern*' Arctic

The most dramatic paleoenvironmental change captured in the ACEX record is the shift from freshwater influenced biosiliceous and organic rich deposits of the Paleogene to fossil poor glaciomarine silty clays that occurred by at least the late early Miocene [22,24]. Based on the original ACEX age model, the timing of this change from euxinic to well-oxygenated open marine conditions was correlated to the tectonically controlled widening of the Fram Strait in the late early Miocene (~17.5 Ma), which allowed a critical two-way surface exchange between the Arctic Ocean and Norwegian Greenland Seas to commence [25].

The transition between these two states is captured in a 5.76 m sequence of cross-banded sediments, interpreted as representing alternating states of oxygenation [26] (Figure 5). This sequence occurs immediately above the 26 Myr hiatus. The origin of the hiatus has been ascribed to either the exposure of the ridge, which may have remained in a shallow water setting through most of the Paleogene [27], or to erosion from enhanced bottom water currents [28], potentially associated with the gradual opening of the Fram Strait [29].

Recent osmium isotope dates from the cross-banded and underlying Eocene age biosiliceous rich sediments suggest that the transition from euxinic to well-oxygenated conditions may have occurred in the late Eocene [23]. Within this new proposed stratigraphy, the cross-banded sediments are interpreted as a condensed section arising from sea-level variations during the opening of the Fram Strait. However, the early age (late Eocene -Oligocene) for a connection between the Atlantic and

Arctic needs to be reconciled with plate tectonic reconstructions, which suggest crustal overlap between the modern Greenland and Svalbard margins during chron 13 time (33.3 Ma) [30], implying that the pre- and synrift topography in this region becomes quite critical to constrain.

Despite emerging questions over the timing and mechanism associated with this profound oceanographic change, from a lithostratigraphic perspective it marks the initiation of *modern* conditions in the central Arctic Ocean. These are defined by the apprent continuous deposition of glaciomarine silts and clays, with a minor but persistent sand component, and containing very limited

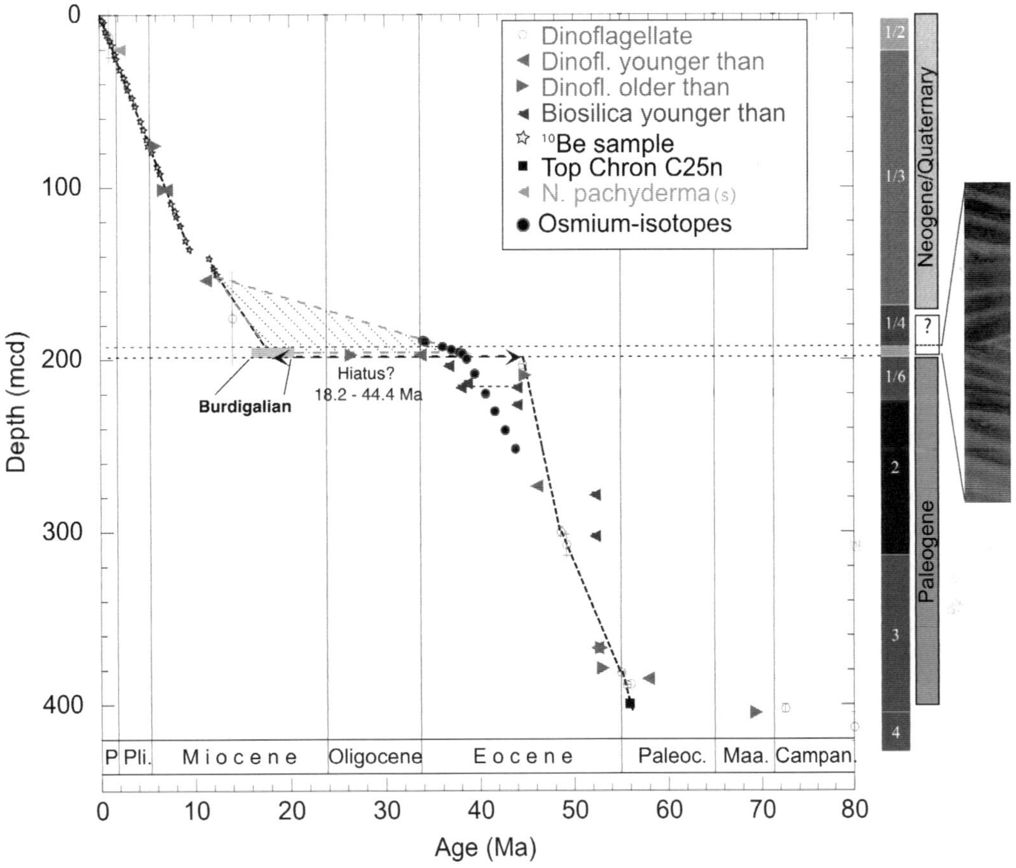

Figure 5. Age model for the ACEX record and corresponding lithologic units [22]. New osmium derived dates [23] are also shown and highlight an emerging discrepency. On the far right is an image of the cross-banded sequence that marks the transition form euxinic to oxygenated bottom water conditions

and sporadic preserved microfossils.

4. Onset and persistence of perennial ice

Based on the results from ACEX there is strong evidence indicating the existence of seasonal ice in the Eocene (47 Ma) [31] when global atmospheric CO_2 concentrations were likely >1000 ppmv. While no direct proxy for perennial ice has been derived from the microfossil poor glaciomarine sediments deposited since at least the late early Miocene, novel solutions for identifying its onset were derived from analyzing the provenance (source region) of clay, heavy mineral and detrital iron-oxide grains found within this record [32,33].

For example, Krylov *et al.* [32] document a pronounced shift in the clay and heavy mineral assemblages (from smectite to illite; and clinopyroxenes to hornblendes respectively) that they ascribe to a change from the Western Laptev/Kara Seas, to the East Siberian Sea as the dominant sediment source area. As the average modern drift time for sea ice to reach the central Lomonosov Ridge from the East Siberian Sea is > 1 year [34], this change is interpreted as marking the onset of perennial ice. This transition is independently dated between 12-14 Ma [22]. Employing similar arguments, Darby [33] uses an iron oxide fingerprinting technique to match detrital iron-oxide grains to a large database of Arctic source regions, and reports a constant input of IRD from all the circum-arctic shelves in samples extending back to at least 15 Ma, again implying that perennial ice has been present throughout this time, and potentially earlier. His analysis also illustrates that tectonic reconstructions from the middle Miocene to present do not significantly alter the distances between the Lomonosov Ridge and Western Arctic and East Siberian Shelves, and cannot be used to explain the changes in sediment provenance in the ACEX record. Frank et al [35] also argue that the generally smooth but coarse resolution exponential decrease of $^{10}Be/^{9}Be$ with depth, and the low calculated flux rates of ^{10}Be into the sediments, provide ancillary evidence for a persistent perennial ice pack since at least 12 Ma.

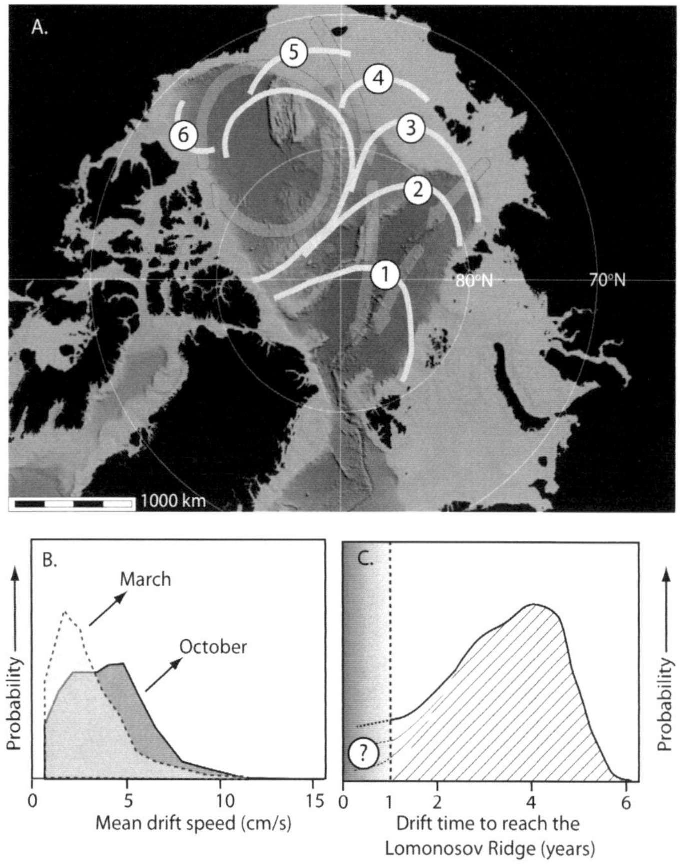

Figure 6. A. Map of the Arctic Ocean showing the dominant surface circulation patterns, and the mean drift times required for ice from different regions to exit through the Fram Strait (adapted from [34]). **B.** Seasonal variations in mean sea ice drift speed in the central Arctic Ocean (generalized and adapted from [44]). **C.** A <u>conceptual</u> illustration of how the mean residence time derived from the mean drift speed may not capture important variability in the statistical distribution. How probable ice drift reaching the Lomonosov Ridge in <1 year (marked by question mark in C) has not yet been quantified for the different source regions.

Within the resolution of the ACEX age model, the onset for perennial Arctic sea ice inferred by Krylov *et al.* [32] occurs at the end of the Middle Miocene Climate Optimum (MMCO, 15-17 Ma), during a period of recognized global cooling [36], reduction in atmospheric CO_2 [37] and the expansion of the East Antarctic ice sheet [38]. It also appears, to a first order, synchronous with terrestrial evidence that documents middle Miocene cooling in the Canadian Arctic Archipelago (CAA) and throughout Alaska and the western North West Territories of Canada [37]. However, the assertion that perennial ice was established after the MMCO and persisted since this time is not consistent with the presence of mixed hardwood and deciduous forests which are preserved in the late Miocene/Pliocene Beaufort Formation of the CAA [40, 41], with early Pliocene sea surface temperature estimates as high as 18°C in the Fram Strait [42] or the early Pliocene peat deposits from Ellesmere Island (78°N, 82°W) recording mean annual temperatures that were ~19°C higher than today [43].

The drift-speed/provenance argument for delineating changes between seasonal and perennial sea ice is simple and somewhat elegant, yet the importance of the findings using this technique (namely that perennial sea ice has been a stable component of the central Arctic Ocean for 12-15 Ma) and the mismatch with other circum-arctic studies, requires further study. For example, [32, 33] both use mean annual residence time charts for sea ice emanating from different regions of the Arctic Ocean. The charts are based upon the monthly analysis of the mean field of sea ice motion derived from drifting buoys deployed on ice flows between 1979 and 1998 as part of the International Arctic Buoy Programme (IABP) [34] and from a paleoceanographic perspective may not capture two important aspects of the drift speed field. First, there is yet no analysis of the statistical distribution of these mean residence times, or quantification of the seasonal, inter-annual and decadal variability associated with them. Although this cannot be resolved in the low-resolution sedimentary records, it clearly defines what is averaged within any one sample over the few thousand years it represents (Figure 6).

A second obstacle is how the modern speeds would be affected by diminished sea ice thickness that would exist during warmer periods of the geologic past. Modern observations suggest a strong link between mean drift speed and sea ice thickness. This was recently illustrated using IABP data [45] in which sea ice drift speeds during the last 29 years increased by 17%/decade in the winter and 85%/ decade in the summer, with no corresponding increase in the derived monthly mean wind speed. This increased drift speed was associated with contemporaneous and observed thinning of the ice pack during this time [46].

At the same time, the arguments that perennial sea ice could not have existed because of the fossilized floral remains along the CAA require a more thorough analysis of the regional climatic effects that would be expected in the Miocene. For example, much warmer than modern conditions along the Northern Greenland coast and the CAA are largely reported from Miocene paleoclimate models, and are associated with a regional anomaly arising from the absence of the Greenland icesheet [47]. In this sense, the ACEX results raise important questions on the controls and overall sensitivity of sea ice in the geologic past, and provide a framework for testing and furthering our understanding of these questions. Due to the burial depth of Miocene age sediments, addressing them fully requires additional deep boreholes from strategic locations in the Arctic.

5. Intensification of Northern Hemisphere glaciations

The growth of large Northern Hemisphere icesheets is widely believed to have occurred between ~2.9-2.6 Ma and is best illustrated by compilations of global benthic $\delta^{18}O$ [47] and by the gradual increase in ice rafted material (IRD) in records from high latitude settings [40,49].

During the last two glacial cycles, prominent fluxes in sand sized material to the Lomonosov Ridge (Figure 3b) are closely related to known changes in the extension of the Barents/Kara icesheet to the shelf edge [13]. These results fit with the growing awareness of glaciogenic bedforms on many topographic highs, imaged with high-resolution subbottom profiling and swath-bathymetry, which indicate the incursion of deep drafting ice (> 900 m draft) into the central Arctic Ocean and collectively suggest the growth of large ice shelves in the late Quaternary [50].

However, while there seems to be a strong correspondence between very subtle changes in grain size and glacial/interglacial cycles through the Quaternary, the ACEX record reveals an overall decrease in the amount of IRD between ~2 Ma (an interval with poor chronostratigraphic control in the ACEX record) and the better-constrained base of Marine Isotope Stage 6 (191 ka) [51] (Figure 7, Figure 4). This reduction in coarse grained IRD cannot be explained by dilution due to higher

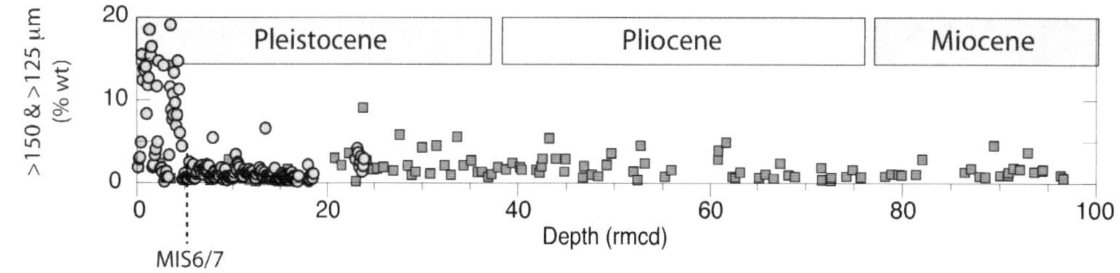

Figure 7. Coarse-fraction content of the ACEX record in the later part of the Cenozoic. >150 µm (squares) and >125 µm (circles). Boundaries of epochs derived from age model of Backman *et al.* [22] that uses an average sedimentation rate of 14.5 m/Myr over this interval. Note the increase in coarse fraction content above the MIS6/7 boundary (see Figure 4 for details) and the low abundances for the middle part of the Pleistocene. No systematic increase in the coarse fraction contents is seen as NH glaciations intensified through the Pleistocene.

sedimentation rates [51], and is not consistent with patterns from other high latitude records that reveal enhanced ice rafting during the Quaternary (*see compilation in* [40]). Existing data from the Lomonosov Ridge indicate that episodes of intense coarse grained ice rafting occur mainly during the last two glacial cycles, which largely corresponds to the ages of dated glaciogenic bedforms [50]. These observations suggest that such massive intrusions of icebergs and ice shelves may not have occurred prior to MIS6, when IRD contents are low in the ACEX record, and furthermore suggest that sea ice was likely the dominant carrier of IRD in previous glacial periods. This remains somewhat speculative, and there is clearly evidence for large-scale ice sheet growth in the Barents Sea during the

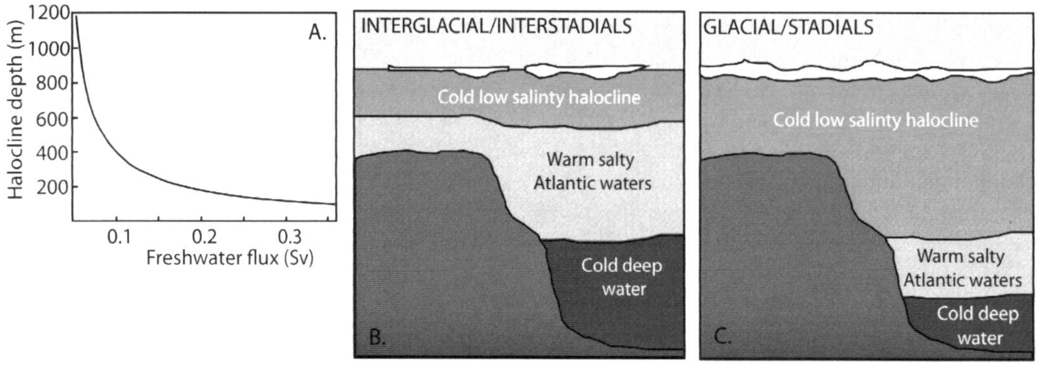

Figure 8. A. Analytically derived changes in the depth to the halocline as a function of freshwater flux to the Arctic Ocean (Jakobsson et al, 2010). **B** and **C** are conceptual illustrations of how the depth to the halocline (and subsequently any warm Atlantic water that enters the Arctic), may be depressed during glacial/stadial periods or at times in the geologic past when freshwater inflow to the Arctic was reduced

middle part of the Pleistocene [49,52]. Longer and better dated sequences from the Amerasian Basin, Morris Jesup Rise and other regions of the Arctic Ocean are needed to address spatial and temporal differences in depositional patterns.

The reduction in the amount of IRD at ~ 2 Ma in the ACEX record coincides with the start of high amplitude oscillations in the neodynium isotopic composition of intermediate waters, which are interpreted as marking the onset of large scale Atlantic water inflow following the subsidence of the Greenland-Scotland Ridge [53]. The observed reductions in IRD that appear coincident with the enhanced inflow of Atlantic water are somewhat counter-intuitive, given the recognized influence that North Atlantic waters have on transporting heat into the central Arctic Ocean. While possible explanations for these observations may lie in the evolution of the major oceanic gateways (the Fram Strait/Barents Sea and the Bering Strait) [51], the role of changing freshwater delivery to the Arctic, long considered to be a fundamental boundary condition for sea ice growth and a possible key to Northern Hemisphere icesheet growth, remain largely unexplored. For example, Jakobsson *et al.* [50] illustrate how the depth of the modern halocline in the Arctic Ocean is extremely sensitive to changes in the freshwater flux, and increases sharply when freshwater delivery is reduced (Figure 8). This suggests that large-scale vertical migration of water masses may occur on glacial/interglacial timescales, and potentially during reduced/enhanced hydrologic periods of the past. To investigate these changes it is critical to exploit coupled analyses of cores that form depth transects in the modern Arctic (*i.e.* Figure 4b). Quantifying these changes and coupling them with proxies of past sea ice and glacial ice, would be of great benefit in attempting to reconstruct the paleoceanographic evolution of the Arctic Ocean, and how it is influenced by oceanic exchanges with the Pacific and Atlantic Oceans.

6. Concluding remarks

Our knowledge of past and present conditions in the Arctic Ocean is advancing rapidly, due in part to the challenge we face in improving our ability to anticipate and predict the magnitude and impacts associated with the rapidly diminishing sea ice cover. From a geologic perspective, an understanding of modern processes provides important groundtruthing for directing our studies into the past and the interpretations we extract from them. The groundbreaking Arctic Coring Expedition proved that scientific drilling in the Arctic Ocean is possible, and it is clear that more long-term records need to be recovered from the Arctic to complement and test theories that have evolved from ACEX. The overview provided here is based primarily on published results from ACEX, and attempts to highlight important questions that are emerging from these results. In doing so it admittedly overlooks many valuable contributions to our understanding of late Cenozoic paleoceanographic change in the Arctic that have emerged from other expeditions and programs. While sea ice is a dominant theme in much of this discussion, it is intimately tied with variations in both terrestrial and oceanic boundary conditions, all of which ultimately need to be integrated into a coherent picture of the geologic and paleoceanographic evolution of this Ocean, and its role in the global climate system.

References

[1] Jakobsson, M 2002 Hypsometry and volume of the Arctic Ocean and its constituent seas *Geochem. Geophy. Geosy.* **3** 1-18

[2] Comiso J C , Parkinson C L , Gersten R and Stock L 2008 Accelerated decline in the Arctic sea ice cover *Geophys Res Lett* **35** doi: 101029/2007GL031972

Serreze M C, *et al* 2007 Perspectives on the Arctic's shrinking sea ice cover *Science* **315** 1533-1536

[3] IPCC 2007 *Climate Change 2007, The physical science basis Contribution of Working Group I to the Fourth Assessment Report of the Intergovernmental Panel on Climate Change* Cambridge University Press, Cambridge and New York

[4] Zhang X and Walsh J E 2006 Toward a seasonally ice-covered Arctic Ocean: scenarios from the IPCC AR4 model simulations *J. Clim.* **19** 1730–1747

Stroeve J, *et al* 2007 Arctic sea ice decline: Faster than forecast *Geophys. Res. Lett.* **34** L09501 doi: 101029/2007GL029703

Wang M and Overland J E 2009 A sea ice free summer Arctic within 30 years? *Geophys. Res. Lett.* **36** L07502

[5] Holland M M and Bitz C M 2003 Polar amplification of climate change in coupled models *Clim. Dynam.* **21** 221-232
[6] Miller G H, Alle R B, Brigham-Grette J, Fitzpatrick J J, Polyak L, Serreze M C, and White, J W C, 2010 Arctic amplification: can the past constrain the future? *Quat. Sci. Rev.* **29**(15-16)1779– 1790
[7] Johannessen O M 2008 Decreasing Arctic sea ice mirrors increasing CO_2 on decadal time scales *Atmos. and Ocean. Sci. Lett.* **1** 51-56
[8] Stroeve J, and Meier W 1999 Updated current year Sea Ice Trends and Climatologies from SMMR and SSM/I, June to September 2001 Boulder, CO: National Snow and Ice Data Center Digital media
[9] Pagani M, Liu Z, LaRiviere J and Ravelo A C 2009 High Earth-system climate sensitivity determined from Pliocene carbon dioxide concentrations, *Nat. Geosci.* **3**, 27-30
[10] de Vernal A, Hillaire-Marcel C, Solignac S, Radi T, and Rochon A 2008 Reconstructing sea ice conditions in the Arctic and sub-Arctic prior to human observations *Geoph. Monograph.* **180** 27-45
[11] Cronin T M, Gemery L, Briggs Jr W M, Jakobsson M, Polyak L and Brouwers E M 2010 Quaternary sea-ice history in the Arctic Ocean based on new Ostracode sea-ice proxy *Quat. Sci. Rev* **29**, 3415-3429
Nørgaard-Pedersen N, Mikkelsen N, Kristoffersen Y, 2007 Arctic Ocean record of last two glacial -interglacial cycles off North Greenland/Ellesmere Island – implications for glacial history *Marine Geology* **244** 93–108
[12] Belt S T, Masse G, Rowland S J, Poulin M, Michel C and LeBlanc B 2007 A novel chemical fossil of palaeo sea ice: IP25 *Organic Gechemistry* **38** 16–27
Muller J, Masse G, Stein R and Belt S T 2009 Variability of sea-ice conditions in the Fram Strait over the past 30,000 years *Nat. Geosci.* **2** 772-776
[13] Spielhagen R F, Baumann K H, Erlenkeuser H, Nowaczyke N R, Nørgaard-Pedersen N, Vogt C and Weiel D 2004 Arctic Ocean deep-sea record of northern Eurasian ice sheet history *Quat. Sci. Rev.* **23** 1455-1483
O'Regan M, *et al.* 2008 Constraints on the Pleistocene chronology of sediments from the Lomonosov Ridge *Paleoceanography* **23** PA1S19 doi:101029/2007PA001551
Sellén E, O'Regan M and Jakobsson M 2010 Spatial and temporal Arctic Ocean depositional regimes: a key to the evolution of ice drift and current patterns *Quat. Sci. Rev.* **29**(25-26) 3644-3664 doi:101016/jquascirev201006005
Polyak L, Bischof J, Ortiz J, Darby D, Channell J, Xuan C, Kaufman D, Lovlie R, Schneider D, Adler R, 2009 Late Quaternary stratigraphy and sedimentation patterns in the western Arctic Ocean. *Global Planet. Change* **68** 5–17
[14] Jakobsson M, Macnab R, Mayer M, Anderson R, Edwards M, Hatzky J, Schenke H-W and Johnson P 2008 An improved bathymetric portrayal of the Arctic Ocean: Implications for ocean modeling and geological, geophysical and oceanographic analyses *Geophys. Res. Lett.* **35** L07602
[15] Cronin T, Eynaud F, Smith S, O'Regan M and King J 2008 Quaternary Paleoceanography of the Central Arctic based on IODP ACEX 302 Foraminiferal Assemblages *Paleoceanography* **23** PA1S18 doi:101029/2007PA001484
[16] Jakobsson M, Løvlie R, Al-Hanbali H, Arnold E, Backman J and Mörth M 2000 Manganese and color cycles in Arctic Ocean sediments constrain Pleistocene chronology *Geology* **28**, 23–26
Jakobsson M, Løvlie R, Arnold E, Backman J, Polyak L, Knudsen J and Musatov E 2001 Pleistocene stratigraphy and paleoenvironmental variation from Lomonosov Ridge sediments, central Arctic Ocean *Global and Planetary Change* **31** 1–22
Kaufman D S, Polyak L, Adler R, Channell J E T and Xuan C 2008 Dating late Quaternary planktonic foraminifer Neogloboquadrina pachyderma from the Arctic Ocean using amino acid racemization *Paleoceanography* **23** PA3224 doi:101029/2008PA001618
[17] Jakobsson M, Backman J, Murray and Løvlie R 2003 Optically stimulated luminescence dating supports central Arctic Ocean cm-scale sedimentation rates *Geochem. Geophy. Geosy.* **4(2)** doi:101029/ 2002GC000423

[18] Hanslik D, Jakobsson M, Backman J, Björck S, Sellén E, O'Regan M, Fornaciari E and Skog G 2010 Pleistocene Arctic Ocean sea ice and deep water isolation times *Quat. Sci. Rev.* **29(25-26)**, 3430-3441 doi:101016/jquascirev201006011

[19] Stein R 2008 *Arctic Ocean Sediments: Processes, Proxies, and Paleoenvrionment* (Elsevier, Amsertdam) p 592

[20] Backman J and Moran K 2009 Expanding the Cenozoic Paleoceanographic record in the Central Arctic Ocean: IODP Expedition 302 Synthesis *Central European Journal of Geosciences* **1(2)** 157-175

[21] Moran K, *et al.* 2006 The Cenozoic palaeoenvironment of the Arctic Ocean *Nature* **441** 601–605

[22] Backman J, et al 2008 Age model and core–seismic integration for the Cenozoic Arctic Coring Expedition sediments from the Lomonosov Ridge *Paleoceanography* **23** PA1S03 doi: 101029/2007PA001476

[23] Poirier A and Hillaire-Marcel C 2009 Os-isotope insights into major environmental changes of the Arctic Ocean during the Cenozoic *Geophys. Res. Lett.* **36** L11602

[24] Stein R, Boucsein B and Meyer H 2006 Anoxia and high primary production in the Paleogene central Arctic Ocean: First detailed records from the LR *Geophys. Res. Lett.* **33**

[25] Jakobsson M, *et al.* 2007 The early Miocene onset of a ventilated circulation regimen in the Arctic Ocean *Nature* **447** 986–990 doi:101038/nature05924

[26] Sangiorgi F, *et al.* 2008 A 25 million year gap in the central Arctic record at the Greenhouse-Icehouse transition: Looking for clues *Paleoceanography* **23** PA1S04 doi: 101029/2007PA001477

[27] O'Regan M, *et al.* 2008 Mid-Cenozoic tectonic and paleoenvironmental setting of the central Arctic Ocean, Paleoceanography **23** PA1S20 doi:101029/2007PA001559

[28] Moore T C and the Expedition 302 Scientists 2006 Sedimentation and subsidence history of the Lomonosov Ridge In Backman J, Moran K, McInroy D B, Mayer L A and the Expedition 302 Scientists Proc IODP 302: Edinburgh (Integrated Ocean Drilling Program Management International, Inc) doi:102204/iodpproc3021052006

[29] Bruvoll V, Kristoffersen Y, Coakley B J and Hopper J R 2010 Hemipelagic deposits on the Mendeleev and northwestern Alpha submarine Ridges in the Arctic Ocean: acoustic stratigraphy, depositional envrionment and an inter-ridge correlation calibrated by the ACEX results *Mar. Geophys. Res.* **31** 149-171 doi: 101007/s11001-010-9094-9

[30] Engen Ø, Faleide J I and Dyreng T K 2008 Opening of the Fram Strait gateway: A review of plate tectonic constraints *Tectonophysics* **450** 51-69

[31] St John K 2008 Cenozoic ice-rafting history of the central Arctic Ocean: Terrigenous sands on the Lomonosov Ridge *Paleoceanography* **23** PA1S05 doi:101029/2007PA001483

Stickley C E, *et al.* 2009 Evidence for middle Eocene Arctic sea ice from diatoms and ice-rafted debris *Nature* **460** 376-379 doi:101038/nature08163

[32] Krylov A A, *et al.* 2008 A shift in heavy and clay mineral provenance indicates a middle Miocene onset of a perennial sea ice cover in the Arctic Ocean *Paleoceanography* **23** PA1S06 doi: 101029/2007PA001497

[33] Darby D A 2008 Arctic perennial ice cover over the last 14 million years *Paleoceanography* **23** PA1S07 doi:101029/2007PA001479

[34] Rigor I G, Wallace J M and Colony R L 2002 Response of Sea Ice to the Arctic Oscillation *J Climate* **15** 2648 - 2668

[35] Frank M, *et al.* 2008 Beryllium isotopes in central Arctic Ocean sediments over the past 12 million years: Stratigraphic and paleoclimatic implications *Paleoceanography* **23** PA1S02 doi: 101029/2007PA001478, pp 1–12

[36] Shevenell A E, Kennett J P and Lea D W 2004 Middle Miocene Southern Ocean cooling and Antarctic cryosphere expansion *Science* **305** 1766–1770 doi:101126/science1100061

[37] Kürschner W M, Kvacek Z and Dilcher D L 2008 The impact of Miocene atmospheric carbon dioxide fluctuations on climate and the evolution of terrestrial ecosystems *PNAS* **105(2)** 449-453

[38] Pekar S F and DeConto R M 2006 High-resolution ice-volume estimates for the early Miocene: Evidence for a dynamic ice sheet in Antarctica *Palaeogeogr. Palaeoclimatol. Palaeoecol.* **231** 101–109

[39] Wolfe J A 1994 An analysis of Neogene climates in Beringia *Palaeogeogr. Palaeoclimatol. Palaeoecol* . **108** 207-216

Fyles J G, Hills L V, Mathews Jr J V, Barendregt R W, Baker J, Irving E and Jette H 1994 Ballast Brook and Beaufort formations (Late Tertiary) on northern Banks Island, Arctic Canada *Quaternary International* **22/23** 141–171

White J M *et al.* 1997 An 18-million-year record of vegetation and climate change in northwestern Canada and Alaska: tectonic and global climatic correlates *Palaeogeogr. Palaeoclimatol. Palaeoecol.* **130** 293–306

[40] Polyak L, *et al.* 2010 History of sea ice in the Arctic *Quat. Sci. Rev.* **29(15-16)** 1757-1778 doi: 101016/jquascirev201002010

[41] Miller G H, *et al.* 2010 Temperature and precipitation history of the Arctic *Quat. Sci. Rev.* **29(15-16)** 1679-171

[42] Robinson M M 2009 New quantitative evidence of extreme warmth in the Pliocene Arctic *Stratigraphy* **6(4)** 265-275

[43] Ballantyne A P, Greenwood D R, Sinninghe Damste J S, Csank A Z, Eberle J J and Rybczynski N 2010 Singnifcantly warmer Arctic surface temperatures during the Pliocene indicated by multiple independent proxies *Geology* **38(7)** 603-606

[44] Martin T and Gerdes R. 2007 Sea ice drift variability in Arctic Ocean Model Intercomparison Project models and observations *Journal of Geophysical Research* **112** C04S10

[45] Rampal P, *et al.* 2009 Positive trend in the mean speed and deformation rate of Arctic sea ice, 1979-2007 *JGR* **114** C05013

[46] Kowk R and Rothrock D A 2009 Decline in Arctic sea ice thickness from submarine and ICESat records: 1958-2008 *Geophys. Res. Lett.* **36** L15501 doi:101029/2009GL039035

[47] Steppuhn A, Micheels A, Bruch A A, Uhl D, Utescher T and Mosbrugger V 2007 The sensitivity of ECHAM4/ML to a double CO2 scenario for the Late Miocene and the comparison to terrestrial proxy data *Global and Planetary Change* **57** 189-212

Tong J A, You Y, Muller R D and Seton M 2009 Climate model sensitivity to atmospheric CO2 concentrations for the Middle Miocene *Global and Planetary Change* **67(3-4)** 129-140

You Y, Huber M, Muller R D, Poulsen C J and Ribbe J 2009 Simulation of the Middle Miocene Climate Optimum *Geophys. Res. Lett.* **36** L04702 doi:101029/2008GL036571

[48] Lisiecki L E and Raymo M E 2005 A Pliocene-Pleistocene stack of 57 globally distributed benthic δ18O record *Paleoceanography* **20** PA1003 doi:101029/2004PA001071

[49] Knies J, *et al.* 2009 The Plio-Pleistocene glaciation of the Barents Sea-Svalbard region: a new model based on revised chronostratigraphy *Quat. Sci, Rev.* **28(9-10)** 812-829

[50] Jakobsson M, *et al.* 2010 An Arctic Ocean iceshelf during MIS 6 constrained by new geophysical and geological data *Quat. Sci. Rev.* **29(25-26)** 3505-3517 doi:101016/ jquascirev201003015

[51] O'Regan M, St John K, Moran K, Backman K, King J, Haley B A, Jakobsson M, Frank M and Röhl U 2010 Plio-Pleistocene trends in ice rafted debris on the Lomonosov Ridge *Quaternary International* **219** 168-176 doi:101016/jquaint200908010

[52] O'Regan M, Jakobsson M, Kirchner N 2010 Glacial geological implications of overconsolidated sediments on the Lomonosov Ridge and Yermak Plateau *Quat. Sci. Rev.* **29(25-26)** 3532-3544 doi:101016/jquascirev201009009

[53] Haley A B, Frank M, Spielhagen R F and Eisenhauer A 2008 Influence of brine formation on Arctic Ocean circulation over the past 15 million years *Nat. Geosci.* **1** 68-72

Marine geology of the St. Lawrence Estuary

Guillaume St-Onge

Canada Research Chair in Marine Geology, Institut des sciences de la mer de Rimouski (ISMER) & GEOTOP Research Center, 310 allée des Ursulines, Rimouski, Québec, Canada, G5L 3A1

College of Oceanic and Atmospheric Sciences (COAS), Oregon State University, 104 Ocean Administration Building, Corvallis, Oregon, USA, 97331

E-mail: guillaume_st-onge@uqar.qc.ca

Mathieu J. Duchesne

Geological Survey of Canada, Québec Division, 490 de la Couronne, Québec City, Québec, Canada, G1K 9A9

Patrick Lajeunesse

Département de géographie & Centre d'études nordiques, Université Laval, Québec City, Québec, Canada, G1V 0A6

Abstract. The St. Lawrence Estuary, Eastern Canada, contains a very thick (>450 m) Quaternary sedimentary sequence. The results from recently conducted geophysical surveys in conjunction with piston coring indicate that these sediments were deposited under very high sedimentation rates, sometimes as high as ~30 m/ka during the last deglaciation. Results also reveal evidence of large submarine landslides during the Holocene, changes in sedimentation rates and the significant role of submarine canyons and channels to transfer sediments from the coast to the deeper marine environment. Finally, this paper highlights the presence of more than 1900 pockmarks on the seafloor of the St. Lawrence Estuary and discusses their possible origins: active hydrocarbon seeps in the Laurentian Channel and biogenic gas seepage on the northwestern shoulder of the Laurentian Channel.

1. Introduction

During the last few years, a significant and fruitful collaboration between several universities, the Geological Survey of Canada and the Canadian Hydrographic Service allowed the acquisition of > 8000 km^2 of multibeam bathymetric data, ~14 000 km of seismic profiles and 700 m of sediment cores from the St. Lawrence Estuary, Eastern Canada (Fig. 1). This new high-resolution database allowed novel observations and discoveries that build upon the pioneer marine geology research conducted by the Geological Survey of Canada in the St. Lawrence Estuary and adjacent Gulf of St. Lawrence in the 1980's and 1990's (e.g., [1-4]). In this paper, we will review some of these recent findings and discuss their implications for stratigraphy, sedimentation, natural hazards and pockmark genesis in the St. Lawrence Estuary. The oceanography and Quaternary geology of the St. Lawrence Estuary and Saguenay Fjord are summarized in this volume in a companion paper [5].

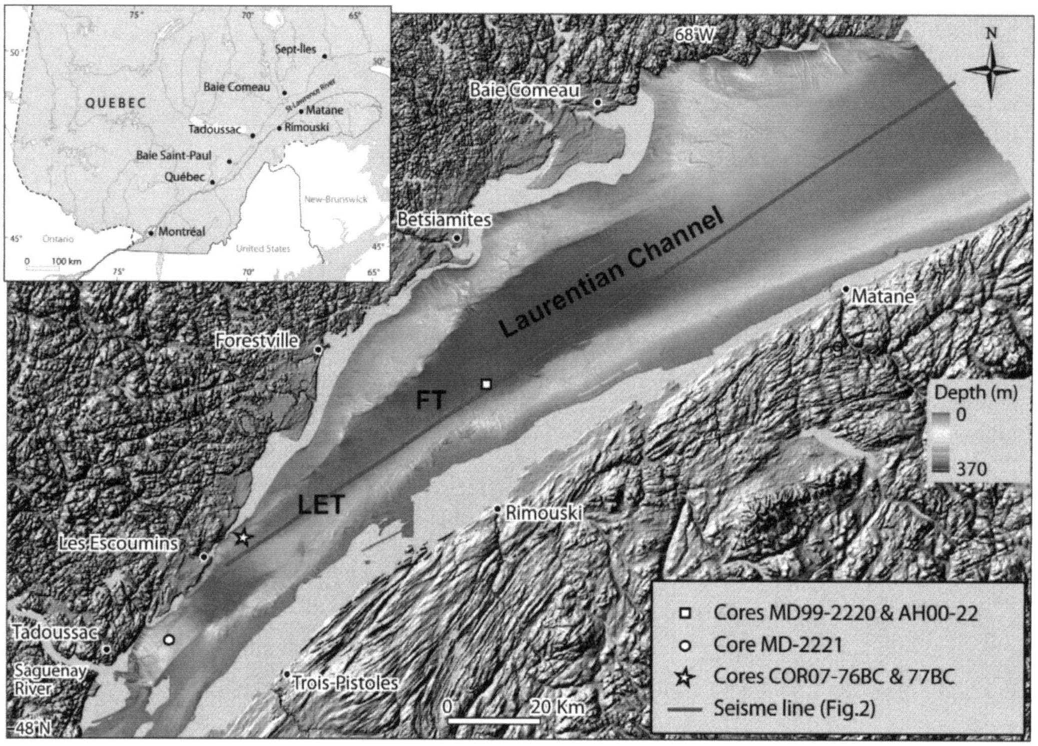

Figure 1. Location of the Lower St. Lawrence Estuary, Eastern Canada. Also illustrated are the multibeam bathymetry (in background), the localities and cores mentioned in the text, the Laurentian Channel, the seismic line presented in Fig. 2 and the location of the Les Escoumins (LET) and Forestville troughs (FT; see text for details and Fig. 5).

2. Stratigraphy of the Quaternary deposits in the St. Lawrence Estuary

High-resolution seismic reflection surveys conducted in the Lower St. Lawrence Estuary allowed the identification of eight seismic units, with a total sediment thickness of more than 450 m in certain areas [6]. Seismic units 1 to 5 are recognized throughout most of the estuary and are illustrated in Figure 2. Units 1 and 2 have a highly variable thickness and fill most of two major bedrock depressions that exist in the Lower St. Lawrence Estuary. Unit 3 is observed over the entire Lower St. Lawrence Estuary basin, partly infilling basins found on the bordering shelves of the estuary. Units 4 and 5 are characterized by a uniform thickness, whereas units 6, 7 and 8 were deposited by local sedimentary processes associated with submarine fans, mass wasting, and contourites along the flanks of the Laurentian Channel [6]. For example, a significant submarine fan is observed offshore the Manicouagan River ([17]; see seismic unit 6 in Figure 5).

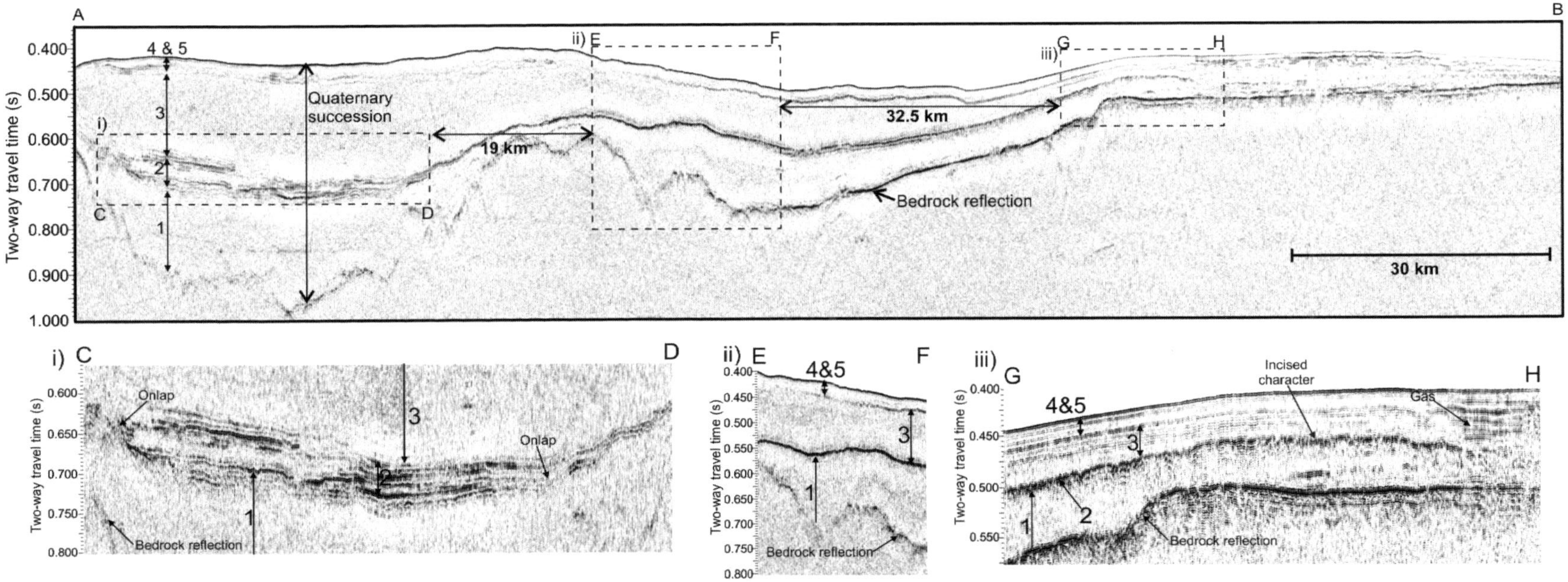

Figure 2. Along-strike seismic stratigraphy of the Lower St. Lawrence Estuary. Illustrated are seismic units 1 to 5 (see text for details). Note the presence of seismic unit 2 throughout the profile. The top of this unit can be used as a chronostratigraphic marker (Younger Dryas). See Figure 1 for location of the seismic line. Modified from [7].

Seismic unit 1 is of unknown age, whereas stratigraphic work indicates that the upper part of seismic unit 2 can be associated with a stilltstand or a local readvance of the Laurentide Ice Sheet (LIS) in the Goldthwait Sea, the proglacial sea that invaded the St. Lawrence Estuary and Gulf following deglaciation (e.g., [8-9]), at or slightly before the Younger Dryas cold interval (~12900 to 11500 cal BP or ~11100 to 10000 ^{14}C yr BP; e.g., [10]). The coring and dating of seismic unit 2 has major implications for the understanding of the thick sedimentary sequence as it indicates that all the sediments above this chronostratigraphic marker were deposited after the Younger Dryas. In some areas, this marker reveals that more that 250 m of sediments were deposited following that event. These very high sedimentations rates can be explained by the proximity of the LIS during deglaciation, where its meltwaters were channelled into the St. Lawrence River system, allowing the deposition of a thick sequence of sediments in the St. Lawrence Estuary. Two long calypso piston cores (MD99-2220 et MD99-2221) sampled in the St. Lawrence Estuary attest to sedimentation rates as high as 30 m/ka during the deglaciation (Fig. 3), but also to a drastic reduction of these sedimentary inputs at 8500 cal BP (7700 ^{14}C yr BP; Figs 3 and 4). This drastic change in sedimentation rates is associated with the re-routing of LIS meltwaters from the St. Lawrence River system to Hudson Bay and Strait following the last outburst flood of Lake Agassiz-Ojibway at 8470 cal BP (~7700 ^{14}C yr BP; [11-12]). According to Clark et al. [13], the St. Lawrence River runoff dropped from ~0.17 Sv to 0.075 Sv (1 Sv = 10^6 m^3s^{-1}) following the re-routing of the meltwaters. The transition from seismic unit 3 to 4 thus marks the drastic change from glaciomarine to postglacial conditions after lake Agassiz-Ojibway final outburst flood [14]. Seismic units 4 and 5 consist of postglacial sediments. This continuous sequence can be disrupted by important mass wasting events (seismic unit 7) that can be observed either directly on the seafloor morphology or on seismic profiles. Some of these submarine landslides were sampled and consist of debris flows, slumps, turbidites or blocky landslides (see section 5).

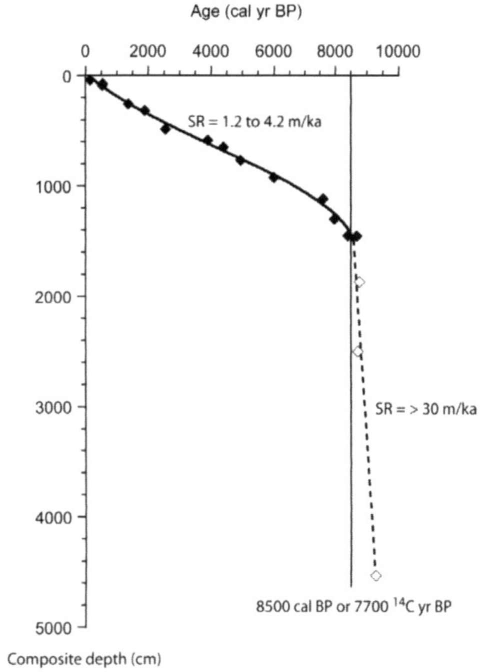

Figure 3. Age-depth relationship for core MD99-2220 (see Fig. 1 for location). This figure illustrates the sharp transition in sedimentation rates at 8500 cal BP (black vertical line). SR=Sedimentation rate. Modified from [15].

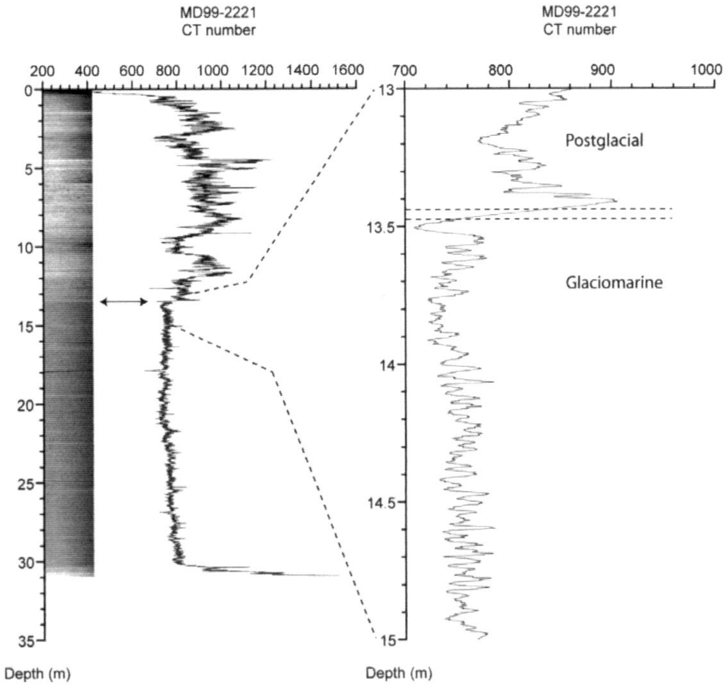

Figure 4. Glaciomarine/postglacial drastic transition observed in core MD99-2221 (see Fig. 1 for location). The left image and the derived CT number were obtained from CAT-scan analysis. Lighter color and higher CT numbers reflect denser (coarser) sediments. Reprinted from [15], with permission from Elsevier.

As mentioned above, seismic unit 1 was never cored in the St. Lawrence Estuary. The upper part of seismic unit 2 was sampled and consists of ice-proximal glaciomarine sediments associated with the Younger Dryas overlain by glaciomarine ice-distal and postglacial sediments. The glaciomarine facies in the St. Lawrence Estuary is characterized by a coarser, sandier proximal facies incorporating ice rafted debris and a finer facies characterized by faintly laminated, massive or plastic silty clays or clayey silts [7, 14, 16]. These glaciomarine sediments are grey to dark grey, but lighter in color than the overlying postglacial sediments. Seismic units 4 and 5 mostly consist of dark grey bioturbated silty clays to sandy muds in the Laurentian Channel and coarser material on the shelves. The stratigraphy of the Lower St. Lawrence Estuary summarized in the model shown in Figure 5 reveals significant sediment accumulation in two troughs identified in the Laurentian Channel: Les Escoumins (LET) and the Forestville (FT) troughs. This figure also highlights the seaward thinning of the sedimentary sequence and the presence of several large submarine landslides (seismic unit 7; see also section 5). The seaward thinning of the sequence thickness indicates that the two troughs acted as important sediment traps during the Quaternary.

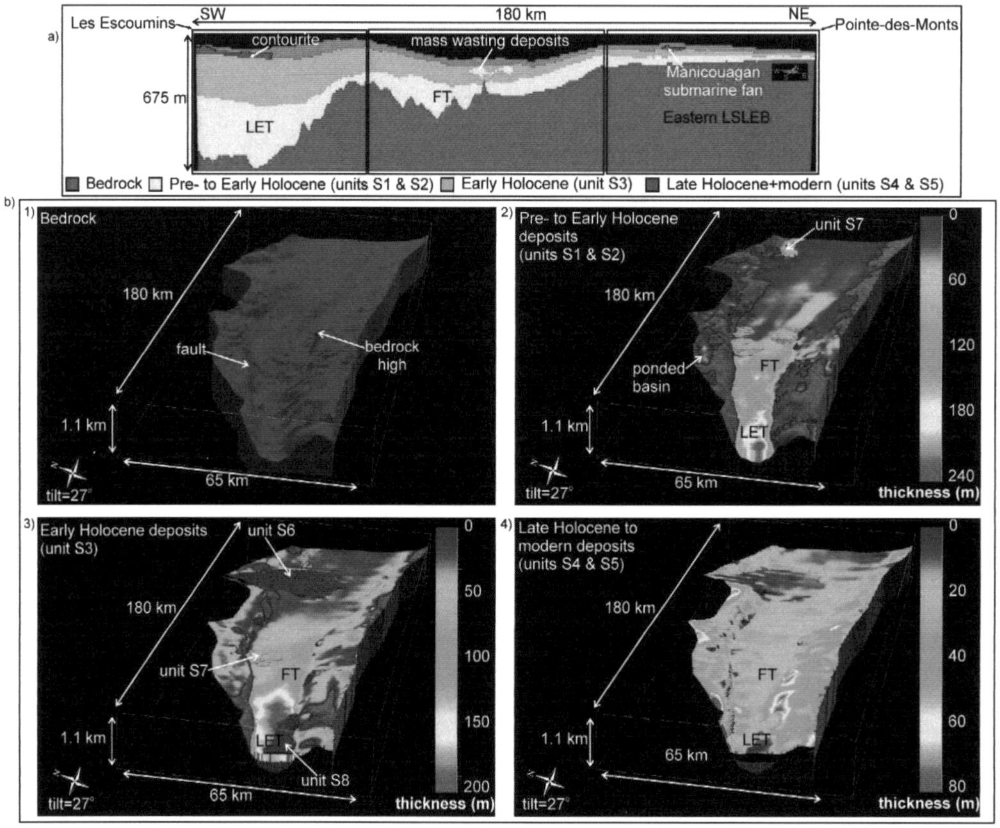

Figure 5. Summary of the stratigraphy and sediment architecture of the Lower St. Lawrence Estuary. Note the significant depocenters in the Les Escoumins (LET) and Forestville (FT) troughs, as well as the presence of several mass wasting events (seismic unit 7) and a significant submarine fan (seismic unit 6). Reprinted from [17], with permission from John Wiley and Sons.

3. High-resolution Holocene magnetostratigraphy

The high sedimentation rates (1.2 to 30 m/ka) and the relatively fine material observed in the postglacial sediments of the Laurentian Channel allowed the reconstruction of the Earth's geomagnetic field variability during most of the Holocene [14]. Using a suite of cores collected in the Laurentian Channel from its head to its mouth, Barletta et al. [16] constructed paleomagnetic master curves based on changes in inclination, declination and relative paleointensity. These master curves were used to determine the high frequency geomagnetic field behaviour in Eastern Canada, but they also have the potential to be used for detailed Holocene magnetostratigraphy and dating purposes.

4. Modern sedimentation in the St. Lawrence Estuary

The seaward thinning of the sedimentary sequence is also expressed in modern sedimentation rates as they diminish exponentially from about 0.74 cm/yr at the head of the Lower St. Lawrence Estuary to 0.04 cm/yr in the Gulf of St. Lawrence to about 0.01 cm/yr at the mouth of the Laurentian Channel in the Atlantic [14, 18, 19, 39]. In addition, major changes in the sedimentation rates are observed in the Lower St. Lawrence Estuary from the Late Holocene to the present, with sedimentation rates increasing from 0.15 to 0.28 cm/yr and finally to 0.74 cm/yr (Fig. 6). These changes are consistent with land-use changes associated with the agricultural development of the European settlement in Eastern Canada, and with the industrialisation, development and growth of the towns along the St. Lawrence River during the second half of the 20th century [14]. The increase in sedimentation rates during the second half of the 20th century coincides with an increase in organic matter fluxes to the seafloor associated with eutrophication due to enhanced nutrient input since the industrialisation [20].

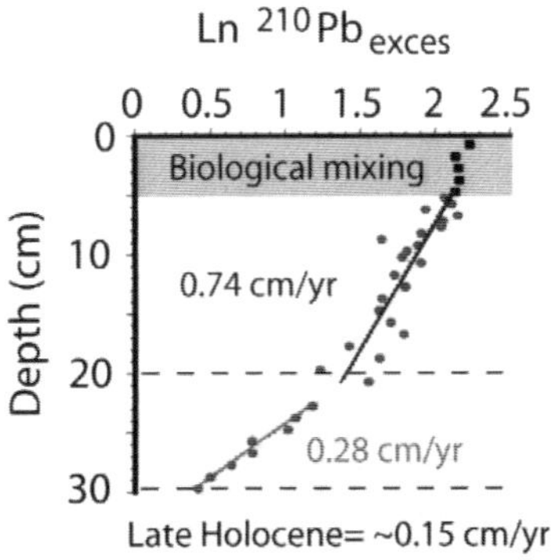

Figure 6. Changes in sedimentation rates derived from ^{210}Pb measurement of core AH00-2220 (see Figure 1 for location). Modified from [14].

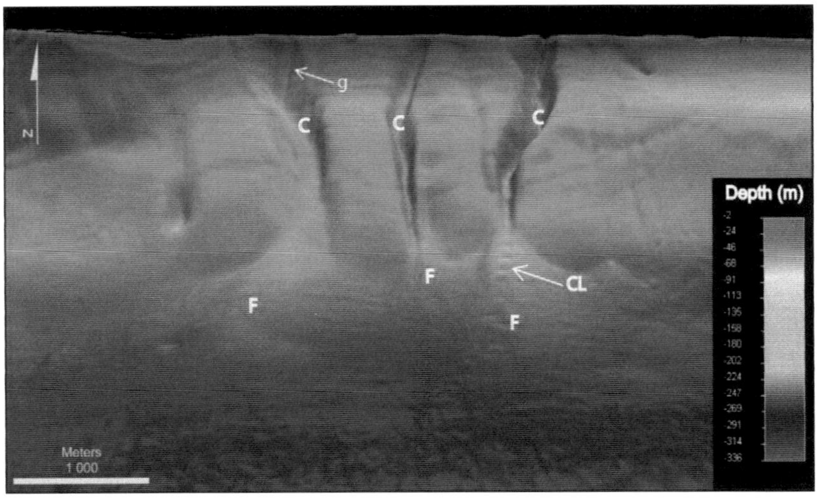

Figure 7. Example of submarine canyons (C) and fans (F) identified along the North Shore of the St. Lawrence Estuary, located East of Les Escoumins. See Fig. 1 for location. A channel-levee (CL) is visible on the fan located on the right side of the image. A small gully (g) is also visible along the headscarp of canyon A. With kind permission from Springer Science+Business Media: [21], © Springer-Verlag 2009.

Modern changes in sedimentation are also observed on the shelf, especially on the North Shore of the St. Lawrence Estuary where coastal erosion significantly increased during the last decades [22]. Multibeam bathymetric data reveal the important role of submarine canyons (Fig. 7) in transferring the eroded coastal sediments from the coast to the bottom of the Laurentian Channel, contributing to the negative sediment budget observed along the coast [21]. Indeed, cores taken in fans at the base of some of these canyons (cores 76 and 77BC, see Fig. 1) revealed a quasi-exponential increase in sand inputs during the last few decades, but also the presence of a turbidite possibly associated with either the 1860 AD Rivière-Ouelle (M~6) or 1870 AD Baie-Saint-Paul (M~6–6.5) earthquakes [23]. Several other fans and channels can be observed on multibeam bathymetric and subbottom profiler data along the North Shore of the St. Lawrence Estuary and can be associated with modern rivers [40]. Similarly, along the North Shore of the Gulf of St. Lawrence, multibeam data collected near the city of Sept-Îles, revealed the presence of a submarine "bird-foot" channel-levee system (Fig. 8). This system most likely reflects significant recent transfer and offshore deposition of coastal sands [24].

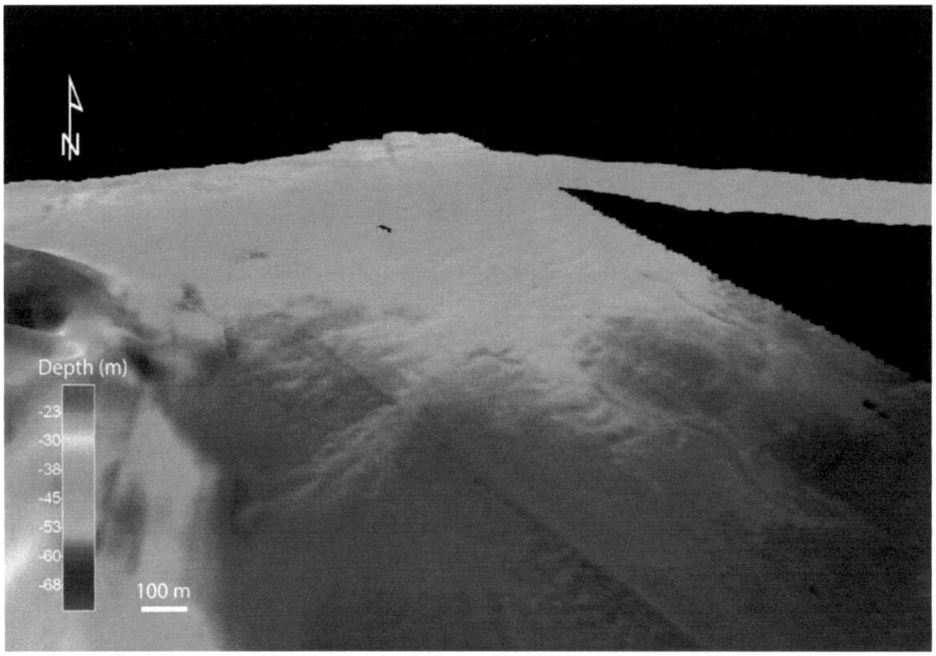

Figure 8. Bird-foot channel-levee system observed offshore the city of Sept-Îles (see inset in Fig. 1 for location).

5. Natural hazards in the St. Lawrence Estuary

The Lower St. Lawrence Estuary is located in the Lower St. Lawrence seismic zone, where about 30 M>2 earthquakes are recorded annually [25]. The largest historical earthquake (M=5.1) recorded in this zone occurred in 1999, about 60 km south of Sept-Îles [25]. The Lower St. Lawrence Estuary is also situated near the most active seismic zone in Eastern Canada, the Charlevoix-Kamouraska seismic zone. During the last 350 years, it was struck by at least 5 earthquakes of magnitude 6 or higher on the Richter scale (e.g., [26]), including the 1663 AD earthquake. The precise epicenter and magnitude of this earthquake are still debated [27], but it likely had a magnitude higher than 7 and occurred either in the Lower St. Lawrence Estuary or in the Saguenay Fjord. A rare combination of high-resolution geophysical, geotechnical and sedimentological terrestrial and marine observations acquired in the

Betsiamites area revealed important terrestrial and submarine features associated to that earthquake, but also to older earthquakes that occurred in the early Holocene [28-29]. Similarly, several seafloor instability features were observed in the Outardes River delta using multibeam bathymetry and were linked to the 1663 AD earthquake [30], whereas a 15-m thick turbidite was observed in the Saguenay Fjord and associated with that event [31]. In addition, Campbell et al. [32] revealed geomorphological and geophysical evidence of Holocene seafloor instabilities features on the southern slope of the Lower St. Lawrence Estuary. These features include lateral spreading, possible liquefaction, rotational slumps, debris flow and blocky submarine landslides. According to Poncet et al. [33], some of these submarine instability features may have been sufficient to generate a tsunami. This is notably the case for a blocky submarine landslide with a 4 km long headwall escarpment near the city of Saint-Siméon (Fig. 9).

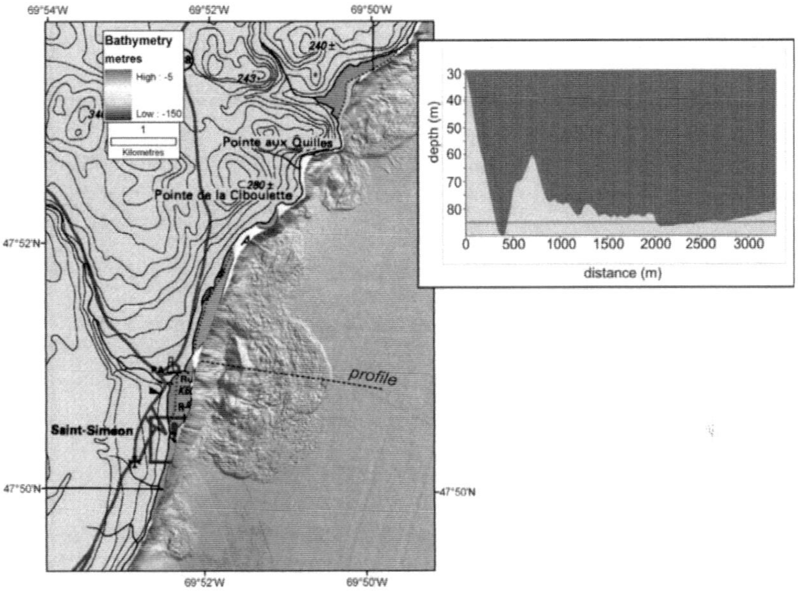

Figure 9. The Saint-Siméon submarine landslide. See Fig. 1 for location. With kind permission from Springer Science+Business Media: [33], © Springer-Verlag + Business Media B.V. 2010.

The presence of these submarine instability features can be explained by a combination of several factors in the St. Lawrence Estuary. As mentioned above, the area is located near the most seismically active zone in Canada. Furthermore, this location experienced extensive glacio-isostatic rebound (e.g., [9]) and very high sedimentation rates following deglaciation. Finally, as illustrated below, the presence of gas in the sediments is also well documented in this sector. All these factors are likely interplaying and favour the generation of submarine landslides in the St. Lawrence Estuary.

6. Pockmarks and gas seeping

According to Lavoie et al. [34], over 1900 pockmarks were recognized in the Lower St. Lawrence Estuary (Fig. 10). In the Matane area alone, more than 109 aligned pockmarks were counted [35]. The pockmarks in the St. Lawrence Estuary generally range from less than 100 m to ~700 m in diameter, whereas the seafloor depression associated with them ranges from a few meters to 25 m deep [34, 36]. They are predominantly observed in the Laurentian Channel and on its northwestern shoulder [36]. On the northwestern shoulder of the Laurentian Channel, high-resolution seismic profiles acquired over pockmark fields reveal some local enhancement of reflections, as well as seismic chimneys that are common below individual pockmarks. These chimneys do not root down to the bedrock. On the other hand, most of the seismic chimneys observed in the Laurentian Channel sediments extend down to the top of the bedrock reflection, suggesting that gas is migrating from the bedrock through the Quaternary sediments and escaping at the seafloor, contributing to the formation of the pockmarks. Underwater video surveys ([37]; **see video (*Animation 1*) in the *Supplementary Data* accompanying this paper**), core sampling and geochemical analyses in some of these pockmarks reveal gas bubbling as well as the presence of carbonate crusts and microbial mats typical of active methane venting submarine environments [34]. These observations led Lavoie et al. [34] to conclude that the pockmarks observed in the Laurentian Channel are the result of active thermogenic gas seeps, while those observed on its northwestern shoulder are likely associated with biogenic gas due to 1) the absence of seismic chimneys rooting down to the top of the bedrock, 2) their association with submarine landslides and 3) possible higher sediment accumulation rates close to organic-rich Quaternary river inputs along the North Shore of the St. Lawrence Estuary [36].

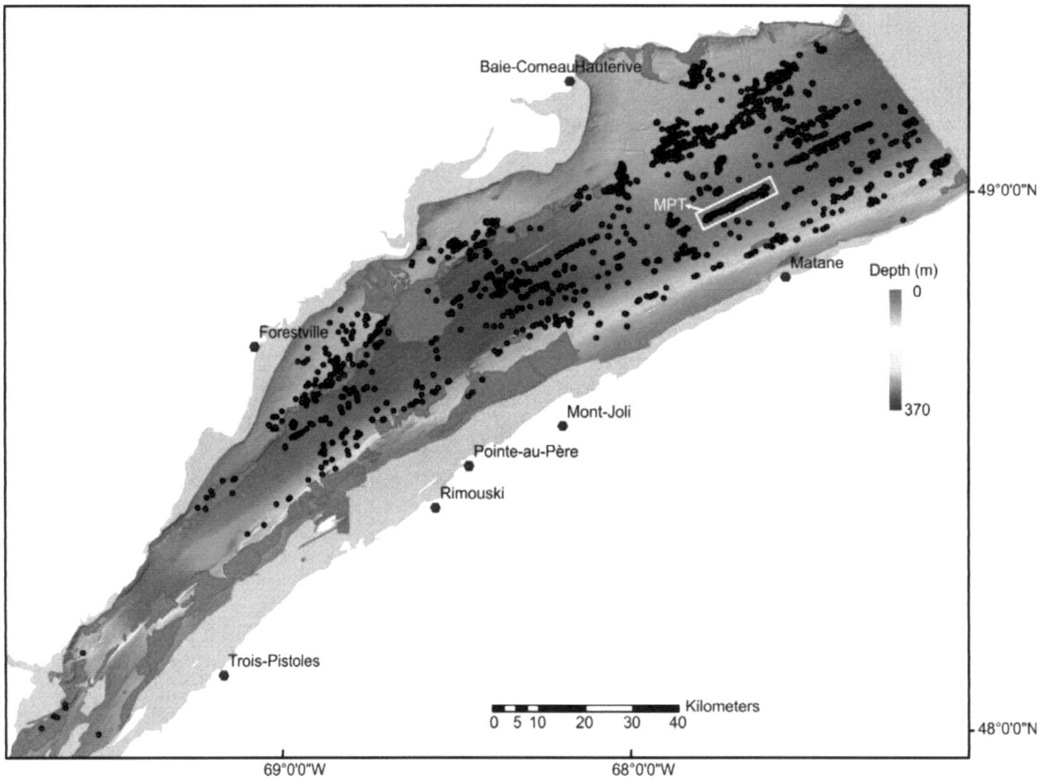

Figure 10. Distribution of pockmarks observed on the Lower St. Lawrence Estuary seafloor. Also illustrated are the Matane pockmark train (MPT) and identified submarine landslides (red polygons). Modified from [38].

7. Conclusions

The findings highlighted in this paper partly details the Quaternary stratigraphy of the Lower St. Lawrence Estuary, a semi-enclosed basin where more than 450 m of sediments were deposited under very high sedimentation rates. These high sedimentation rates allowed the preservation of a continuous Quaternary sediment package and suggest that the St. Lawrence Estuary has a unique potential as a long coring site to study waxing and waning phases of the Laurentide Ice Sheet and long-term environmental changes at an unprecedented resolution. The work summarized in this short paper also illustrates evidence of several significant submarine landslides and their possible association with strong earthquakes, as well as recent changes in sedimentation rates and the potential role of submarine canyons and channels to transfer coastal sediments offshore. Finally, this paper highlights the presence of an important number of pockmarks on the seafloor and the presence of active hydrocarbon seeps in the Laurentian Channel and along its shoulders.

Acknowledgements

The authors sincerely thank the captains, officers, crew and students of the RV Marion Dufresne II, Alcide C. Horth, Coriolis II, CCGS Calanus, CCGS Creed and CCGS Matthew for their help and dedication during several cruises in the St. Lawrence Estuary. This research was supported by the Natural Sciences and Engineering Research Council of Canada (NSERC) through ship time and discovery grants to G.S. and P.L. and by the *Fonds québecois de la recherche sur la nature et les technologies* (FQRNT) through a new professor grant to P.L. Part of this work was also carried out under the Hydrocarbon Potential of the Gulf of St. Lawrence and the Surrounding Area and the Geoscientific Mapping of the St. Lawrence Estuary projects of the Geological Survey of Canada. Special thanks to Jacques Locat (U. Laval) for a continuous and productive collaboration during expeditions and for fruitful discussions on submarine landslides and marine geology of the St. Lawrence Estuary. Similarly, we are in debt to Andrée Bolduc and Nicolas Pinet for fruitful discussions and for their implication in acquiring some of the geophysical data presented in this paper, as well as Jacques Labrie (ISMER) for successfully converting the video file in the appropriate format and size. We also thank Calvin Campbell, Virginia Brake and Andrée Bolduc for their constructive reviews. Finally, we thank Richard Larocque (DFO-IML) and his team for providing video footage of active gas seeps. This is GEOTOP contribution no. 2011-04 and GSC contribution no. 20100413.

References

[1] Syvitski J P M and Praeg D B 1989 Quaternary sedimentation in the St. Lawrence Estuary and adjoining areas, Eastern Canada: An overview based on high-resolution seismo-stratigraphy. *Géographie physique et Quaternaire* **43** 291-310

[2] Syvitski J P M 1992 Marine geology of Baie des Chaleurs. *Géographie physique et Quaternaire* **46** 331-34

[3] Hein F J, Syvitski J P M, Dredge L A and Long B F 1993 Quaternary sedimentation and marine placers along the North Shore, Gulf of St. Lawrence. *Canadian Journal of Earth Sciences* **30** 553–574

[4] Josenhans H and Lehman S 1999 Late glacial history of the Gulf of St. Lawrence, Canada. *Canadian Journal of Earth Sciences* **36** 1327–1345

[5] de Vernal A, St-Onge G and Gilbert D 2011 Oceanography and Quaternary geology and of the St. Lawrence Estuary and the Saguenay Fjord. *IOP Conference Series: Earth and Environmental Science* **this volume**

[6] Duchesne M J, Pinet N, Bolduc A, Bédard K and Lavoie D 2007 Seismic stratigraphy of the Lower St. Lawrence Estuary Quaternary deposits and seismic signatures of the underlying geological domains. Geological Survey of Canada, *Current Research 2007-D2*, 14 pp

[7] St-Onge G, Lajeunesse P, Duchesne M J and Gagné H 2008. Identification and dating of a key Late Pleistocene stratigraphic unit in the St. Lawrence Estuary and Gulf (Eastern Canada). *Quaternary Science Reviews* **27** 2390-2400

[8] Dionne J-C 1977 La mer de Goldthwait au Québec. *Géographie physique et Quaternaire* **31** 61–80

[9] Dionne J-C 2001 Relative sea-level changes in the St. Lawrence estuary from glaciation to present day. *Deglacial History and Relative Sea-Level Changes, Northern New England and Adjacent Canada* (Geological Society of America Special Paper 351) ed T K Weddle and M J Retelle (Boulder: Geological Society of America) pp 271-284

[10] Broecker W S 2006 Was the Younger Dryas triggered by a flood? *Science* **312** 1146-1148

[11] Barber D C, Dyke A, Hillaire-Marcel C, Jennings A E, Andrews J T, Kerwin M W, Bilodeau G, McNeely R, Southon J, Morehead M D and Gagnon J-M 1999 Forcing of the cold event of 8,200 years ago by catastrophic drainage of Laurentide lakes. *Nature* **400** 344-348

[12] Lajeunesse P and St-Onge G 2008. Subglacial origin of Lake Agassiz-Ojibway final outburst flood. *Nature Geoscience* **3** 184-188

[13] Clark P U, Marshall S J, Clarke G K C, Hostetler S W, Licciardi, J M and Teller J T 2001 Freshwater forcing of abrupt climate change during the last glaciation. Science **293** 283-287

[14] St-Onge G, Stoner J S and Hillaire-Marcel C 2003. Holocene paleomagnetic records from the St. Lawrence Estuary: centennial- to millennial-scale geomagnetic modulation of cosmogenic isotopes. *Earth and Planetary Science Letters* **209** 113-130

[15] St-Onge G and Long B 2009 CAT-scan analysis of sedimentary sequences: an ultrahigh-resolution paleoclimatic tool. *Engineering Geology* **103** 127-133

[16] Barletta F, St-Onge G, Stoner J S, Lajeunesse P and Locat J 2010 A high-resolution Holocene paleomagnetic secular variation and relative paleointensity stack from eastern Canada. *Earth and Planetary Science Letters* **298** 162-174

[17] Duchesne M J, Pinet N, Bédard K, St-Onge G, Lajeunesse P, Campbell C and Bolduc A 2010 Role of the bedrock topography in the Quaternary filling of a giant semi-enclosed basin: the Lower St. Lawrence Estuary, Eastern Canada. *Basin Research* **22** 933-951

[18] Muzuka A N N and Hillaire-Marcel C 1999. Organic matter burial rates along the eastern Canadian Margin and stable isotope constraints on its origin and diagenetic evolution. *Marine Geology* **160** 251-270

[19] Smith J N and Schafer C T 1999 Sedimentation, bioturbation, and Hg uptake in the sediments of the estuary and Gulf of St. Lawrence. *Limnology and Oceanography* **44** 207-219

[20] Thibodeau B, de Vernal A and Mucci A 2006 Enhanced primary productivity, organic carbon fluxes and the development of hypoxic bottom waters in the Lower St. Lawrence Estuary, Eastern Canada: micropaleontological and geochemical evidence. *Marine geology* **231** 37-50

[21] Gagné H, Lajeunesse P, St-Onge G and Bolduc A 2009 Transfer of recently eroded coastal sediments to the deeper marine environment through submarine canyon and fan systems: examples from the Lower St. Lawrence Estuary (Eastern Canada). *Geo-Marine Letters* 29 191-200

[22] Bernatchez P, Dubois J-M 2004 Bilan des connaissances de la dynamique de l'érosion des côtes du Québec maritime laurentien. *Géographie physique et Quaternaire* **58** 45-71

[23] Gouin P 2001 *Historical earthquakes felt in Quebec (from 1534 to March 1925)* (Montreal: Guérin)

[24] Lajeunesse P, St-Onge G, Labbé G and Locat J 2007 Multibeam bathymetry and morphostratigraphy of submarine mass-movements offshore Sept-îles, NW Gulf of St. Lawrence. *Submarine Mass Movements and Consequences* ed V Lykousis, D Sakellariou and J Locat (Dordrecht: Springer) pp 287-296

[25] Lamontagne M, Keating P and Perreault S 2003 Seismotectonic characteristics of the Lower St. Lawrence Seismic Zone, Quebec: insights from geology, magnetics, gravity and seismics. *Canadian Journal of Earth Sciences* **40** 317-336

[26] Lamontagne, M. 2009. Possible earthquake triggers of submarine landslides in the estuary of the St. Lawrence River: insights from the earthquake catalogue. Geological Survey of Canada, *Open file 6007* 61 pp

[27] Locat J 2008 Localisation et magnitude du séisme du 5 février 1663 (Québec) revues à l'aide des mouvements de terrain. *Proceedings of the 4th Canadian Conference on Geohazards : From*

Causes to Management ed J Locat, D Perret, D Turmel, D Demers and S Leroueil (Québec: Presse de l'Université Laval) pp 429-443

[28] Cauchon-Voyer G, Locat J and St-Onge G 2008 Late-Quaternary morpho-sedimentology and submarine mass movements of the Betsiamites area, Lower St. Lawrence Estuary, Quebec, Canada. *Marine Geology* **251** 233-252

[29] Cauchon-Voyer G, Locat J, Leroueil S and St-Onge G in press Large-scale subaerial and submarine Holocene and recent mass movements in the Betsiamites area, Quebec, Canada. *Engineering Geology*, doi: 10.1016/j.enggeo.2011.04.011

[30] Duchesne M J, Long B F, Urgeles R and Locat J 2003 New evidence of slope instability in the Outardes Bay delta area, Quebec, Canada. *Geo-Marine Letters* **22** 233-242

[31] St-Onge G, Mulder T, Piper D J W, Hillaire-Marcel C and Stoner J S 2004 Earthquake and flood-induced turbidites in the Saguenay Fjord (Québec): a Holocene paleoseismicity record. *Quaternary Science Reviews* **23** 283-294

[32] Campbell C D, Duchesne, M and Bolduc A 2008 Geomorphological and geophysical evidence of Holocene seafloor instability on the southern slope of the Lower St. Lawrence estuary, Québec. *Proceedings of the 4th Canadian Conference on Geohazards : From Causes to Management* ed J Locat, D Perret, D Turmel, D Demers and S Leroueil (Québec: Presse de l'Université Laval) pp 367-374

[33] Poncet R, Campbell C, Dias F, Locat J and Mosher D 2010 A study of the tsunami effects of two landslides in the St. Lawrence Estuary. *Submarine Mass Movements and Their Consequences* (Advances in Natural and Technological Hazards Research 28) ed D C Mosher, C Shipp, L Moscardelli, J Chaytor, C Baxter, H Lee and R Urgeles (Dordrecht: Springer) pp 755-764

[34] Lavoie D, Pinet N, Duchesne M, Bolduc A and Larocque R 2010 Methane-derived authigenic carbonates from active hydrocarbon seeps of the St. Lawrence Estuary, Canada. *Marine and Petroleum Geology* **27** 1262-1272

[35] Pinet N, Duchesne M and Lavoie D 2010 Linking a linear pockmark train with a buried Palaezoic structure : a case study from the St. Lawrence Estuary. *Geo-Marine Letters* **30** 517-522

[36] Pinet N, Duchesne M, Lavoie D, Bolduc A and Long B 2008 Surface and subsurface signatures of gas seepage in the St. Lawrence estuary (Canada) : significance to hydrocarbon exploration. *Marine and Petroleum Geology* **25** 271-288

[37] Bolduc, A, Larocque, R and Duchesne, M J 2008 Des évents actifs dans l'estuaire maritime du Saint-Laurent, Québec. Geological Survey of Canada, *Current Research 2008-21*, 8 pp

[38] Pinet N, Brake V, Duchesne, M J and Campbell D C 2010 What's the St. Lawrence Estuary shallow subsurface and seafloor made of? GeoCanada 2010 - Working with the Earth. Meeting Abstracts, 4 pp

[39] Zhang D C 2000 *Fluxes of short-lived radioisotopes in the marginal marine basins of eastern Canada* PhD Thesis Environmental Sciences (Montréal: Université du Québec à Montréal) pp 193

[40] Bolduc A, Bernatchez P and Côté R 2008 Les paysages marins et côtiers de la Côte-Nord québécoise: une vue du large – Excursion à bord du NGCC F.G. Creed. XIth meeting of the *Association québécoise pour l'étude du Quaternaire* (AQQUA). Field Guide, 21 pp

Oceanography and Quaternary geology of the St. Lawrence Estuary and the Saguenay Fjord

Anne de Vernal[1], Guillaume St-Onge[2] and Denis Gilbert[3]

[1] GEOTOP, Université du Québec à Montréal, P.O. Box 8888, succursale "centre ville" Montréal, QC, H3C 3P8 Canada

[2] GEOTOP and Institut des Sciences de la Mer (ISMER), Université du Québec à Rimouski, 310, Allée des Ursulines, Rimouski, QC, G5L 3A1 Canada

[3] Institut Maurice-Lamontagne, Pêches et Océans Canada, 850 route de la Mer, CP 1000 Mont-Joli, QC, G5H 3Z4 Canada

E-mail: devernal.anne@uqam.ca

Abstract. The St. Lawrence Estuary is an environment marked by important freshwater discharge and well stratified water masses, recording large seasonal contrast in surface waters from freezing conditions in winter to temperate conditions in summer due to a very strong seasonal cycle in overlying air temperature. High productivity takes place in the pelagic and benthic environments, where a recent trend toward bottom water hypoxia is observed. The area was profoundly marked by the Quaternary glaciations. Thick glaciomarine sequences dating from the last deglaciation are observed in the Estuary and along the shores, whereas a relatively thin layer (a few meters at most) of hemipelagic mud was deposited during the Holocene.

1. Introduction

The St. Lawrence Estuary is a transitional environment between the St. Lawrence River and the northwest North Atlantic Ocean [1, 2]. It is characterized by strong stratification of water masses. The low salinity surface layer is marked by a large amplitude gradient of temperatures, freezing in winter and reaching up to 18°C in summer. The St. Lawrence Estuary is unique from several points of view. It shows features typical of polar environments despite its geographical location between 47 and 50°N. Seasonal sea-ice cover develops for several months each year and the fauna includes cold water species such as harp seals and beluga whales [3, 4]. Alpine tundra vegetation occurs at moderately low altitudes on the north and south shores [5]. Moreover, the landscape is profoundly marked by past glacial activity. For example, the Laurentian Channel (Figure 1) and the Saguenay Fjord are valleys resulting from glacial erosion [6].

In the present paper, we describe the main oceanographic features of the Estuary and Gulf of St. Lawrence. We briefly discuss the methodological difficulty of conducting paleoceanographic investigations in such an estuarine environment and provide a few references to regional studies. Finally, we present the issue of bottom water hypoxia.

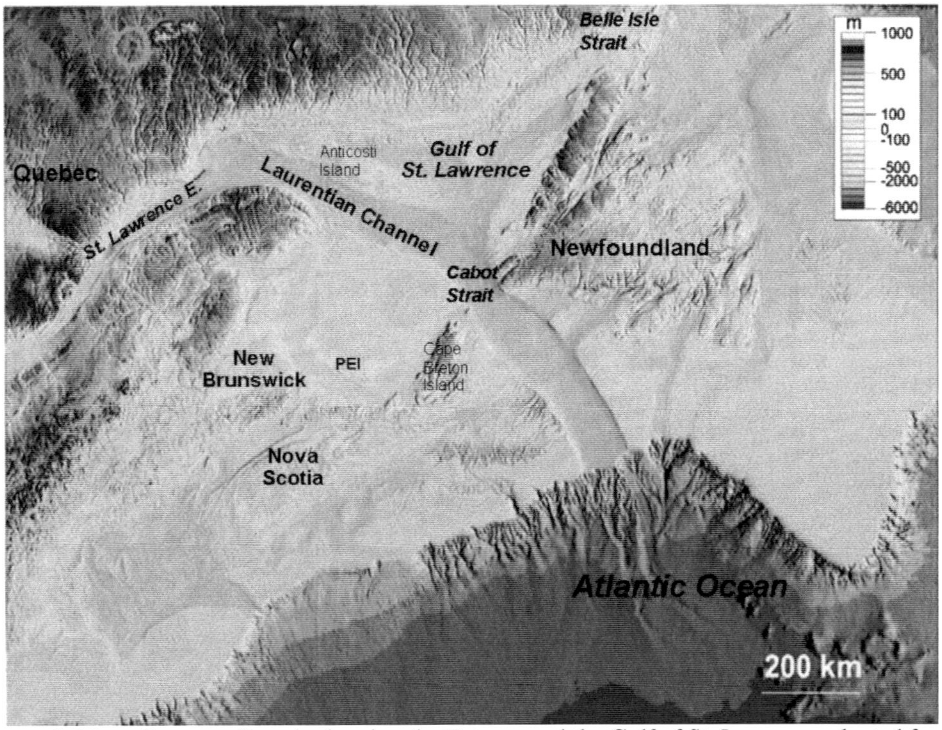

Figure 1. Map of eastern Canada showing the Estuary and the Gulf of St. Lawrence adapted from Shaw et al. [7].

2. Oceanographic context of the Estuary and Gulf of St. Lawrence

The St. Lawrence Estuary and Gulf form together an epicontinental sea making the transition from land to the northwest North Atlantic Ocean. It is characterized by a very large watershed (Figure 2) and a mean freshwater discharge of 10 900 $m^3 s^{-1}$ [8].

The circulation in the Estuary and Gulf of St. Lawrence is typical of estuarine systems and results in strong water masses stratification. The water column extends down to 500 m in the Laurentian Channel. It is characterized by three water layers (Figure 3). The upper one forms a thin surface layer (< 50 m) and flows seaward. It is marked by low salinity (27-31.5) due to the mixing of seawater with the freshwater runoff. As a consequence of stratification, the surface layer is relatively thin [9] and thus characterized by a low thermal inertia: it freezes in winter for up to 3 or 4 months per year and warms up to 18°C in summer. Seasonal gradients of temperature and productivity are thus extremely high [10]. Below the surface water, a cold intermediate layer extends down to about 150 m (Figure 4). This cold intermediate layer is closely related to climate conditions since it originates from winter cooling and water density increase in addition to brine release during sea-ice formation [9, 11]. Below the intermediate layer, warmer (4-6°C) and saltier (~ 34.5) waters occupy the Laurentian Channel. These waters originate from the mixing of the Labrador Current Water (LCW) and the North Atlantic Central Water (NACW) along the continental margin and form the St. Lawrence bottom waters [2]. The bottom waters flow landward, towards the head of the Laurentian Channel in the lower St. Lawrence Estuary near Tadoussac at the mouth of the Saguenay Fjord. The head of the Laurentian Channel is thus characterized by upwelling, which results in extremely high productivity suitable for the development of large whale populations.

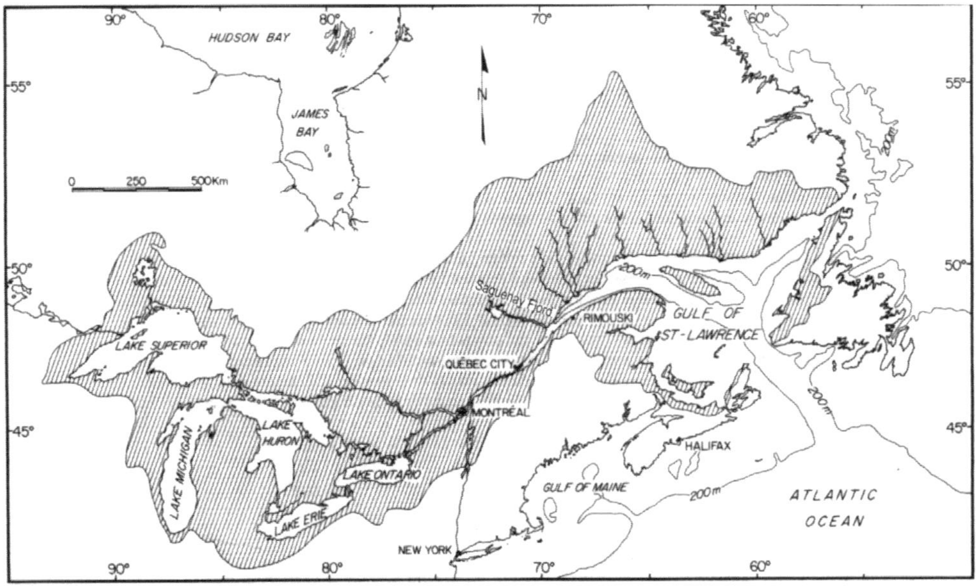

Figure 2. Map of eastern Canada showing the extent of the Gulf of St. Lawrence watershed adapted from Koutikonsky and Bugden [1].

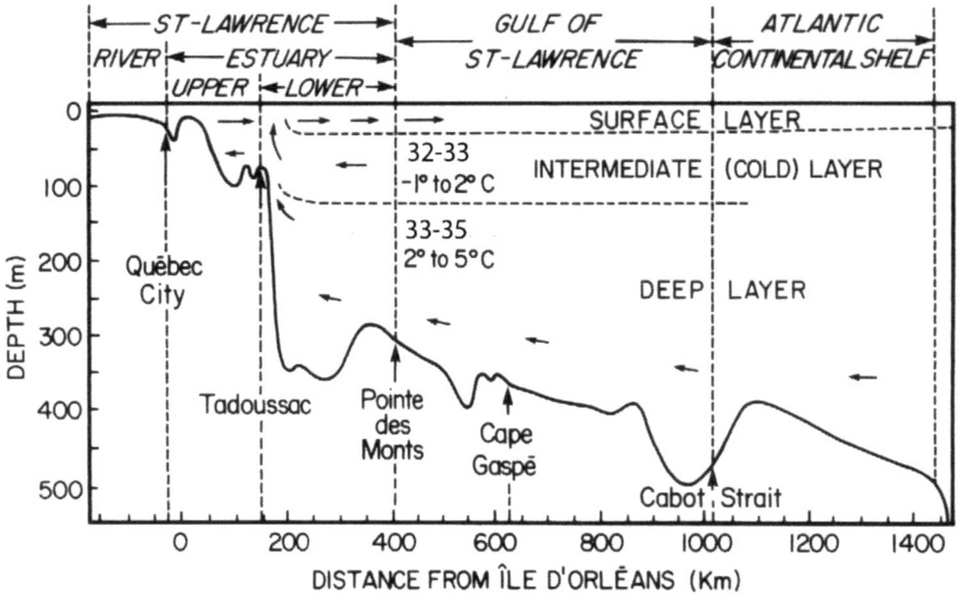

Figure 3. Schematic illustration of water masses in the St. Lawrence system (from the River to the Ocean) adapted from Koutikonsky and Bugden [1].

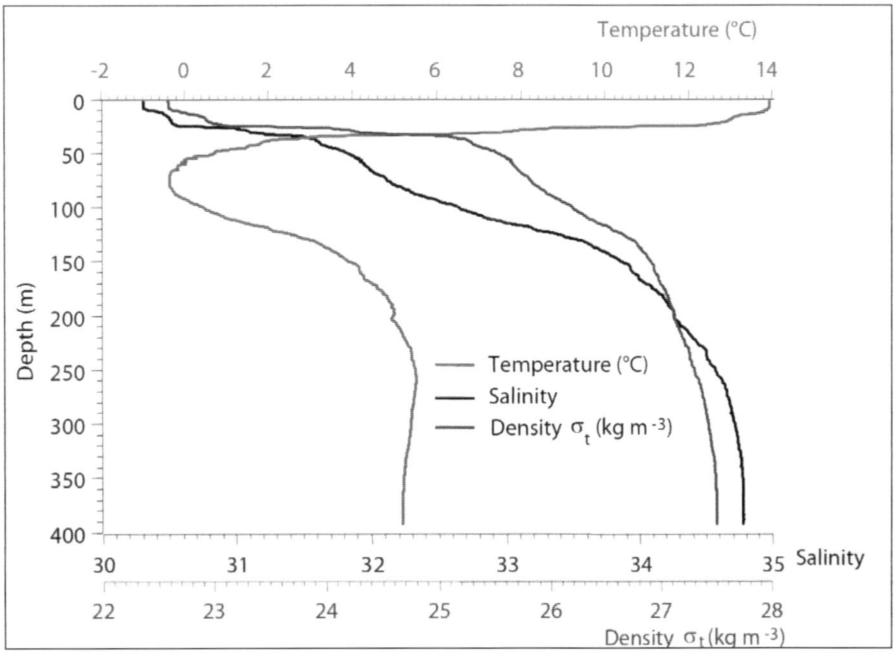

Figure 4. Summer salinity, temperature, and density profile in the Lower St. Lawrence Estuary.

In the St. Lawrence Estuary, the primary productivity is largely dominated by diatoms accompanied by dinoflagellates [12]. Protists forming calcareous shells such as coccolithophorids and planktic foraminifers, which are usually common in surface waters of mid latitude oceans, are rare in the upper water column, probably because of the low surface salinity. Among planktonic foraminifers, *Neogloboquadrina pachyderma* left-coiled is the most common species. The occurrence of this polar species, which is adapted to develop in the mesopelagic layer, appears compatible with the cold conditions of the intermediate and bottom water layers where the salinity range is also suitable. Benthic foraminifers are usually recovered in relatively high numbers in sediments [13]. However, the preservation of calcium carbonate is often an issue because of large organic carbon fluxes [14, 15]. The degradation of organic matter by oxidation results in CO_2 production, thus lowering the pH in the sediment and contributing to calcium carbonate dissolution.

3. Hydrography of the Saguenay Fjord

The Saguenay Fjord joins the Lower St. Lawrence Estuary near Tadoussac. The fjord is almost 80 km long and forms deep basins reaching 275 m of depth. It is separated from the Estuary by a sill about 25 m deep. Because of the large drainage basin with mean annual discharge of about 1500 m^3/s [16], the salinity gradient is particularly sharp. The shallow surface layer (~10 m) is characterized by a salinity of about 0 near the head to 29 at the mouth of the fjord (Figure 5). The surface water freezes in winter and reaches up to 16°C in summer.

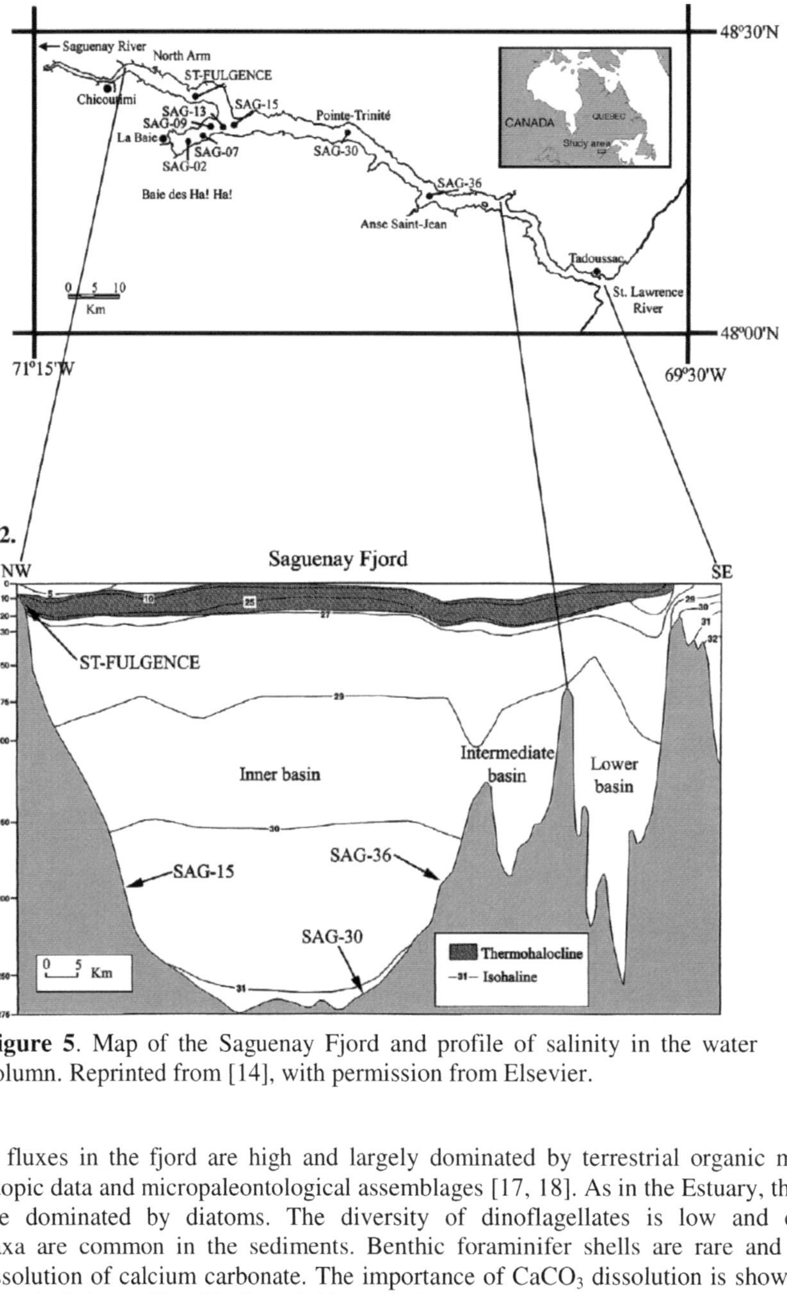

Figure 5. Map of the Saguenay Fjord and profile of salinity in the water column. Reprinted from [14], with permission from Elsevier.

The carbon fluxes in the fjord are high and largely dominated by terrestrial organic matter as shown from isotopic data and micropaleontological assemblages [17, 18]. As in the Estuary, the protist assemblages are dominated by diatoms. The diversity of dinoflagellates is low and cysts of heterotrophic taxa are common in the sediments. Benthic foraminifer shells are rare and there is evidence for dissolution of calcium carbonate. The importance of $CaCO_3$ dissolution is shown by the abundance of organic linings of benthic foraminifers relative to calcareous shells [14].

4. Regional paleoceanography

The paleoceanography of the St. Lawrence Estuary or Saguenay Fjord is difficult to explore for several reasons. First, the glacial activity eroded most sediment that accumulated prior to the last deglaciation. Second, the rarity of biogenic carbonate makes it difficult to develop ^{14}C chronology and the use of $\delta^{18}O$ and $\delta^{13}C$ in foraminifers. Third, the oceanographic conditions are complex due to the estuarine character of the environment, which is marked by large amplitude gradients of temperature and salinity in time and space, both horizontally and vertically. Thus, methodological approaches should be adapted to establish a chronological framework and to make sensible paleoceanographic reconstructions.

For the establishment of chronological schemes, several approaches can be used. At the scale of the last two centuries, ^{210}Pb and ^{137}Cs measurements from core sediments collected in the Laurentian Channel from the Lower St. Lawrence Estuary permit to evaluate sedimentation rates of the order of 0.4-0.5 cm/year and a mixing depth of bioturbation of about 3-4 cm [15, 19]. On longer time scale, high-resolution paleomagnetic data using paleomagnetic secular variations (inclination and declination) and relative paleointensity calibrated against a ^{14}C time scale may be very useful [20]. Moreover, using several radiocarbon-dated cores from the head to the mouth of Laurentian Channel, Barletta et al. [21] constructed a regional paleomagnetic master curve that has the potential to be used for dating purposes. Available data from the Lower St. Lawrence Estuary show sedimentation rates of the order of 0.15 cm/year at the scale of the last 8500 years and much higher sediment accumulation rates (>3 cm/year) before, likely in relation with outwash deposition of glacio-marine and fluvio-glacial material [20]. These high sedimentation rates allowed the deposition of >450 m of sediments in the Laurentian Channel [22], where at least half the sequence was deposited after the Younger Dryas cold episode [23].

For the reconstruction of past sea-surface conditions, diatoms that are abundant in sediment can provide very useful information in the Estuary and Gulf of St. Lawrence [24]. Organic-walled dinoflagellate cysts also are useful tracers [25], whereas benthic foraminifers may help to reconstruct benthic conditions [13-15].

Most published studies on the paleoceanography of the St. Lawrence system are based on cores collected seaward in the Laurentian Channel of the Gulf of St. Lawrence and on the Laurentian Fan. At these locations, sedimentation rates are lower and the cored sediment may span longer than the Holocene. Moreover, because sea-surface salinity is higher, the use of conventional paleoceanographic tools such as the isotopic composition of foraminifers is possible. The published data from the most oceanic parts of the Gulf of St. Lawrence illustrate large amplitude changes from the last ice age to modern. They evidence extremely cold conditions, with more than 6 months/year of sea-ice cover during an interval corresponding to the Younger Dryas episode [26, 27]. They also show that the cold event corresponds to reduced meltwater outflow from the southern Laurentide Ice Sheet through the St. Lawrence Estuary axis [27].

Other studies dealing with the paleoceanography of the St. Lawrence system are based on exposure of postglacial sediments above sea-level dated of about 12 000 to 8500 years BP. During the last deglaciation, the northward retreat of the Laurentide Ice Sheet led to marine transgression in the St. Lawrence Lowlands. The Champlain and the Goldthwait seas then developed West and East of Quebec City respectively. The marine limits of the Champlain Sea reached up to 170 m in the South and 250 m in the North of its extent. Postglacial marine deposits are thus exposed to air at many locations and form terraces along the North and South shores of the St. Lawrence Estuary and throughout the St. Lawrence Lowlands. Micropaleontological and geochemical analyses of sediments from the Champlain Sea suggest environmental conditions and fauna resembling those of the modern Arctic seas during the deglaciation [28, 29, 30].

5. Hypoxia

The St. Lawrence Estuary is characterized by a strong stratification of water masses which prevents oxygen replenishment through mixing and diffusion [31]. In the Laurentian Channel, the bottom waters are isolated from the atmosphere by a permanent pycnocline situated between 100 and 150 m water depth. Under such conditions, the water gradually loses oxygen landward from Cabot Strait to Rimouski (Figure 6) through respiration and remineralization of organic matter that settles through the water column. Measurements of dissolved oxygen in the near bottom waters have shown a significant decrease from ~125 to <65 μmol L^{-1} over the past 75 years, leading to severe and persistent hypoxia [31].

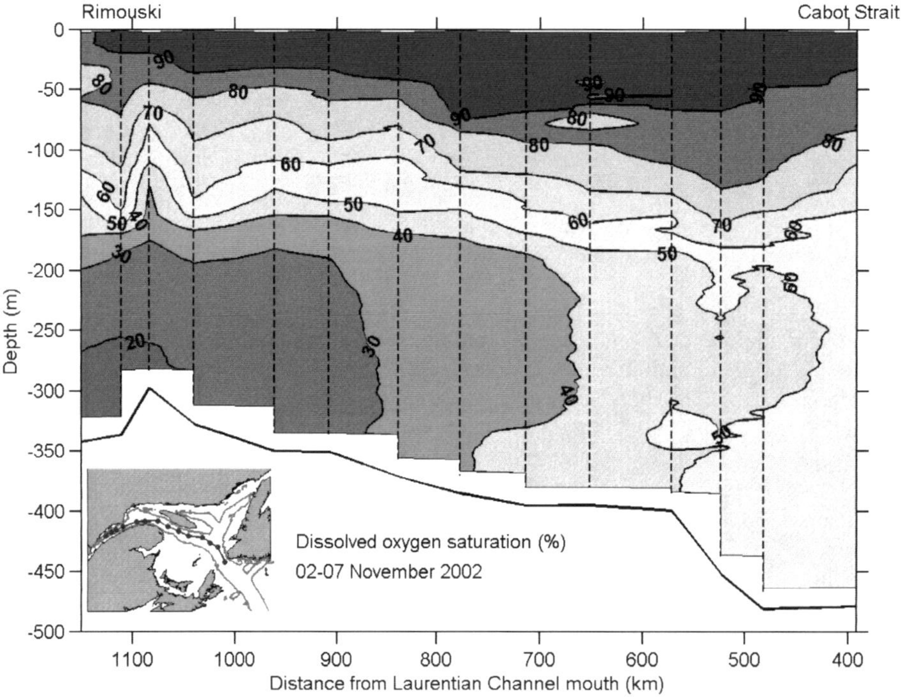

Figure 6. Contours of dissolved oxygen content in the water column of the Estuary and Gulf of St. Lawrence (adapted from [31]).

Two causes are invoked to explain the recent hypoxia. One is related to primary productivity and carbon fluxes leading to oxygen consumption during degradation of the organic matter. Eutrophication linked to enhanced nutrient input since the beginning of the industrial period has been suggested [15]. The other cause invoked is an increase in bottom water temperatures [31]. The authors attributed the bottom water warming to a decrease in the proportion of oxygen rich water from the Labrador Current and an increase in the proportion of oxygen poor water from the North Atlantic subtropical gyre thus explaining between half and two thirds of the oxygen decline in the bottom waters of the Lower St. Lawrence Estuary. In addition, this warming may have accelerated the remineralisation rate of organic matter and oxygen consumption in the water column and sediment [32, 33]. Both factors probably played a determinant role. The remaining questions are the actual part of anthropogenic forcing on eutrophication and bottom water warming and the resilience time of the system. In order to answer such questions, sedimentary cores are currently being analysed with the aim to establish centennial to millennial time series of bottom water conditions and pelagic productivity. Nonetheless, the recent work of Thibodeau et al. [34] using the isotopic composition of benthic foraminifera indicates that the recent trend observed in the bottom temperatures of the Lower St. Lawrence Estuary has no equivalent over the last millennium. Moreover, compound-specific $\delta^{15}N$ of deep-sea gorgonian corals indicates these warm temperatures may be unprecedented in the last 1800 years [35].

6. Conclusion

The St. Lawrence Estuary is a large transitional environment characterized by stratified water masses. Being under the influence of a cold temperate climate, it is marked by sea-ice formation in winter, but warm sea-surface conditions in summer. From this point of view, it is a unique environment, which has been suggested as an analogue for the northern North Atlantic during the last glacial maximum [36]. The St. Lawrence Estuary was marked by glacial erosion during the last ice age. The sedimentary sequences cored in the St. Lawrence Estuary mostly cover the last deglaciation and the Holocene. Above thick late glacial sequences, the postglacial deposits consist in a mixture of terrigenous inputs from the watershed and pelagic fluxes. The St. Lawrence Estuary is thus a particularly complex environment in which the sedimentary sequences are related to the hydrological budgets of a large watershed in addition to inflow from the northwest North Atlantic. Documenting paleoceanographic conditions in this context is a challenge. .

References

[1] Koutitonsky V G and Bugden G L 1991 The physical oceanography of the Gulf of St. Lawrence: a review with emphasis on the synoptic variability of the motion *Can. Spec. Publ. Fish. Aquat. Sci.* **113** 57–90

[2] Dickie L M and Trites W 1983 The Gulf of St. Lawrence *Estuaries and Enclosed Seas* ed H Ketchum (Amsterdam: Elsevier) 403–25

[3] Saucier F J, Roy F, Gilbert D, Pellerin P and Ritchie H 2003 Modeling the formation and circulation processes of water masses and sea ice in the Gulf of St. Lawrence, Canada *J. Geophys. Res.* C **108** 3269 doi:10.1029/2000JC000686

[4] Lesage V, Hammill M O and Kovacs K M 2001 Marine mammals and the community structure of the Estuary and Gulf of St Lawrence, Canada: evidence from stable isotope analysis *Mar. Ecol. Prog. Ser.* **210** 203–21

[5] Rousseau C 1974 *Géographie floristique du Québec-Labrador. Distribution des principales espèces vasculaires* (Québec: Les Presses de l'Université Laval) 799 p.

[6] Dyke A S and Prest V K 1987 Late Wisconsinan and Holocene history of the Laurentide Ice Sheet *Geogr. Phys. Quat.* **41** 237–63

[7] Shaw J, Gareau P and Courtney R C 2002 Palaeogeography of Atlantic Canada 13–0 ky *Quat. Sci. Rev.* **21** 1861–78

[8] Bourgault D and Koutitonsky V G 1999 Real-time monitoring of the freshwater discharge at the head of the St. Lawrence Estuary *Atmosphere - Ocean* **37** 203–20

[9] Craig J and Gilbert D 2008 Estimation of mixed layer depth at the AZMP fixed stations. *AZMP Bulletin PMZA* **7** 37–42

[10] Le Fouest V, Zakardjian B, Saucier, F J and Çizmeli, S 2006 Application of SeaWIFS- and AVHRR-derived data for mesoscale and regional validation of a 3-D high-resolution physical-biological model of the Gulf of St. Lawrence (Canada) *J. Marine Syst.* **60** 30–50

[11] Gilbert D and Pettigrew B 1997 Interannual variability (1948-1994) of the CIL core temperature in the Gulf of St. Lawrence *Can. J. Fish. Aquat. Sci.* **54** 57–67

[12] Therriault J C and Levasseur M 1985 Control of phytoplankton production in the lower St. Lawrence estuary: light and freshwater runoff *Naturaliste Canadien* **112** 77–96

[13] Rodrigues C G and Hooper K 1982 Recent benthonic foraminiferal associations from offshore environments in the Gulf of St. Lawrence *J. Foramin. Res.* **12** 327–52

[14] Leduc J, Bilodeau G, de Vernal A and Mucci A 2002 Distribution of benthic foraminiferal populations in surface sediments of the Saguenay Fjord, before and after the 1996 flood *Paleogeogr. Paleoclimatol. Paleoecol.* **180** 207–23

[15] Thibodeau B, de Vernal A and Mucci A 2006 Enhanced primary productivity, organic carbon fluxes and the development of hypoxic bottom waters in the Lower St. Lawrence Estuary, Eastern Canada: micropaleontological and geochemical evidence *Mar. Geol.* **231** 37–50

[16] Fortin G R and Pelletier M 1995 *Synthèse des connaissances sur les aspects physiques et*

chimiques de l'eau et des sédiments du Saguenay (*Rapport technique, zones d'intervention prioritaire 22–23*) (Montréal: Centre Saint-Laurent, Environnement Canada) 192 p.

[17] St-Onge G, Leduc J, Bilodeau G, de Vernal A, Devillers R, Hillaire-Marcel C, Loucheur V, Marmen S, Mucci A and Zhang D 1999 Caractérisation des sédiments récents du fjord du Saguenay (Québec) à partir de traceurs physiques, géochimiques, isotopiques et micropaléontologiques *Geogr. Phys. Quat.* **53** 339–50

[18] St-Onge G and Hillaire-Marcel C 2001 Isotopic constraints on sedimentary inputs and organic carbon burial rates in the Saguenay Fjord, Quebec *Mar. Geol.* **176** 1–22

[19] Jennane A 1992 *Application de la méthode du Plomb-210 dans l'estuaire et le golfe du St-Laurent* MSc Thesis Université du Québec à Montréal 85 p.

[20] St-Onge G, Stoner J S and Hillaire-Marcel C 2003 Holocene paleomagnetic records from the St. Lawrence Estuary, eastern Canada: Centennial- to millennial-scale geomagnetic modulation of cosmogenic isotopes *Earth Planet. Sci. Lett.* **209** 113–30

[21] Barletta F, St-Onge G, Stoner J S, Lajeunesse P and Locat J 2010 A high-resolution Holocene paleomagnetic secular variation and relative paleointensity stack from eastern Canada *Earth Planet. Sci. Lett.* **298** 162–74

[22] Duchesne M J, Pinet N, Bédard K, St-Onge G, Lajeunesse P, Campbell P and Bolduc C 2010 Role of the bedrock topography in the Quaternary filling of a giant semi-enclosed basin: the Lower St. Lawrence Estuary, Eastern Canada *Basin Research* doi: 10.1111/j.1365-2117.2009.00457

[23] St-Onge G, Lajeunesse P, Duchesne M J and Gagné H 2008 Identification and dating of a key Late Pleistocene stratigraphic unit in the St. Lawrence Estuary and Gulf (Eastern Canada) *Quat. Sci. Rev.* **27** 2390–400

[24] Lapointe M 2000 Modern diatom assemblages in surface sediments from the Maritime Estuary and the Gulf of St. Lawrence, Québec *Mar. Micropaleontol.* **40** 243–65

[25] de Vernal A and Giroux L 1991 Distribution of organic walled microfossils in recent sediments from the Estuary and Gulf of St. Lawrence: some aspects of the organic matter fluxes *Can. J. Fish. Aquat. Sci.* **113** 189–99

[26] de Vernal A, Guiot J and Turon J-L 1993 Postglacial evolution of environments in the Gulf of St. Lawrence: palynological evidence *Geogr. Phys. Quat.* **47** 167–80

[27] de Vernal A, Hillaire-Marcel C and Bilodeau G 1996 Reduced meltwater outflow from the Laurentide ice margin during the Younger Dryas *Nature* **381** 774–77

[28] Hillaire-Marcel C 1981 Paleo-océanographie isotopique des mers post-glaciaires du Québec *Palaeogeogr. Palaeoclimatol. Palaeoecol.* **35** 63–119

[29] Paiement M 2007 *Reconstitution des paléoenvironnements de la mer de Champlain à partir de palynomorphes terrestres et marin* MSc thesis Université du Québec à Montréal 67 p.

[30] Harington C R, Lebel S, Paiement M and de Vernal A 2006 Félix. A Late Pleistocene White whale (*Delphnapterus leucas*) skeleton from Champlain Sea deposits at Saint-Félix-De-Valois-Québec *Geogr. Phys. Quat.* **60** 183–98

[31] Gilbert D, Sundby B, Gobeil C, Mucci A and Tremblay G H 2005 A seventy-two-year record of diminishing deep-water oxygen in the St. Lawrence estuary: The northwest Atlantic connection *Limnol. and Oceanogr.* **50** 1654–66

[32] Genovesi L, de Vernal A, Thibodeau B, Hillaire-Marcel C, Mucci A and Gilbert D Recent changes in bottom water oxygenation and temperature in the Gulf of St. Lawrence: Micropaleontological and geochemical evidence *Limnol. Oceanogr.* in press

[33] Gillooly J F, Brown J H, West G B, Savage V M and Charnov E L 2001 Effects of size and temperature on metabolic rate *Science* **293** 2248–51

[34] Thibodeau B, de Vernal A, Hillaire-Marcel C and Mucci A 2010 20th Century warming in deep waters of the Gulf of St. Lawrence: a unique feature of the last millennium *Geophys. Res. Lett.* **37** L17604 doi: 10.1029/2010GL044771

[35] Sherwood O A, Lehmann M F, Schubert C J, Scott D B and McCarthy M D 2011 Nutrient

regime shift in the western North Atlantic indicated by compound-specific $\delta^{15}N$ of deep-sea gorgonian corals *Proc. Natl Acad. Sci.* **108** 1011–15

[36] de Vernal A, Eynaud F, Henry M, Hillaire-Marcel C, Londeix L, Mangin S, Matthiessen J, Marret F, Radi T, Rochon A, Solignac S and Turon J-L 2005 Reconstruction of sea-surface conditions at middle to high latitudes of the Northern Hemisphere during the Last Glacial Maximum (LGM) based on dinoflagellate cyst assemblages *Quat. Sci. Rev.* **24** 897–924

Planktonic foraminifera in the Arctic: potentials and issues regarding modern and quaternary populations

Frédérique Eynaud

Université Bordeaux I, Laboratoire EPOC (Environnements et Paléoenvironnements OCéaniques), UMR CNRS 5805, Avenue des facultés, 33405 Talence cedex – France

E-mail: f.eynaud@epoc.u-bordeaux1.fr

Abstract. Calcareous microfossils are widely used by paleoceanographers to investigate past sea-surface hydrology. Among these microfossils, planktonic foraminifera are probably the most extensively used tool (e.g. [1] for a review), as they are easy to extract from the sediment and can also be used for coupled geochemical (e.g; $\delta^{18}O$, $\delta^{13}C$, Mg/Ca) and paleo-ecological investigations. Planktonic foraminifera are marine protists, which build a calcareous shell made of several chambers which reflect in their chemistry the properties of the ambient water-masses. Planktonic foraminifera are known to thrive in various habitats, distributed not only along a latitudinal gradient, but also along different water-depth intervals within surface waters (0-1000 m). Regarding their biogeographical distribution, planktonic foraminifera assemblages therefore mirror different water-masses properties, such as temperature, salinity and nutrient content of the surface water in which they live. The investigation of the specific composition of a fossil assemblage (relative abundances) is therefore a way to empirically obtain (paleo)information on past variations of sea-surface hydrological parameters. This paper focuses on the planktonic foraminifera record from the Arctic domain. This polar region records peculiar sea-surface conditions, with the influence of nearly perennial sea-ice cover development. This has strong impact on living foraminifera populations and on the preservation of their shells in the underlying sediments.

1. Introduction

The Arctic region has been identified as one of the most sensitive areas with regard to global environmental changes [2], with estimations of an atmospheric warming from 6.5 to 8°C by 2100 (in reference to the 1980-1999 period). Some evidence [3, 4] suggests an Arctic Ocean almost free of sea-ice in September within the next decades, although most models predict this loss within the next century [2]. Feedbacks and mechanisms behind a comparable evolution are still missing and poorly understood (grouped under the term "Arctic amplification", [5]) but their detection, and hopefully their comprehension, is possible by the study of past climatic analogs. These analogs are recorded in fossil archives. For paleoceanographers, their investigation relies on the recovery of sedimentary sequences with high temporal resolution which provide detailed records of the ocean dynamics (front or current migration) in relation to the climatic system. Such kinds of sequences lie in the deep-sea, containing a large amount of fossilized material among which are the planktonic foraminifera. These small 'bugs" are probably the most basic tool for paleoceanographic and paleohydrological reconstructions of the late Quaternary (e.g. [6, 1]). They are used as material for geochemical analyses (stable isotopes, major elemental ratio, ^{14}C dating...) or studied as (paleo)communities which mirror the

Published under licence by IOP Publishing Ltd

(paleo)environmental parameters of a specific time snapshot (see [1] for a review). This paper briefly synthesizes issues and potentials regarding the use of planktonic foraminifera in sediments of the Arctic Ocean.

2. Arctic planktonic foraminifera assemblages and their paleoenvironmental significance

Boreal regions are characterised by low pelagic carbonate productivity. For modern planktonic foraminifera, it has been estimated to be half the observed standing stocks of tropical or temperate regions [7-10]. Together with the corrosive nature of bottom waters, it explains the poor preservation and scarcity of calcareous remains detected in Arctic sediments. Interpretation of Arctic sedimentary records is thus hampered by this limitation (e.g. [11, 12]).

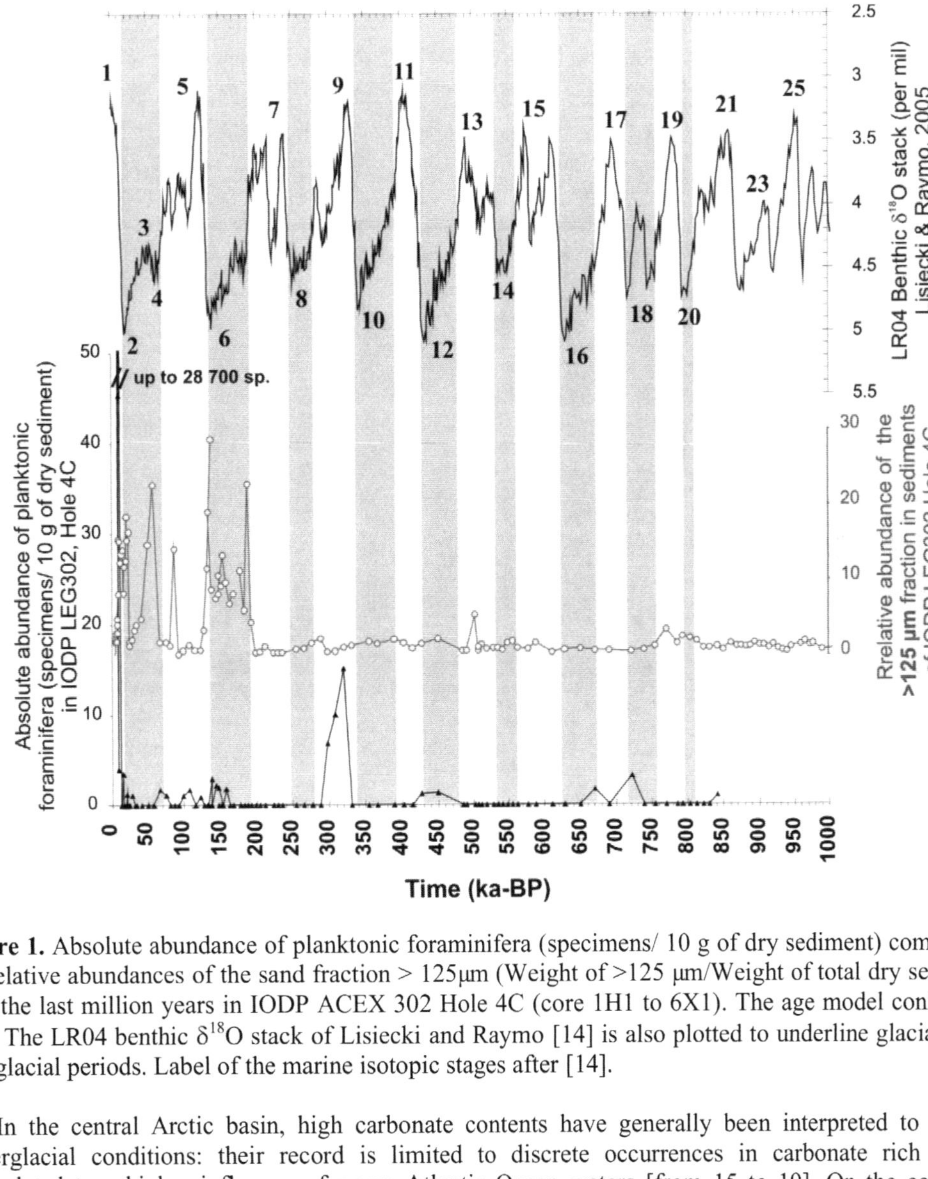

Figure 1. Absolute abundance of planktonic foraminifera (specimens/ 10 g of dry sediment) compared to the relative abundances of the sand fraction > 125µm (Weight of >125 µm/Weight of total dry sed. x100) over the last million years in IODP ACEX 302 Hole 4C (core 1H1 to 6X1). The age model conforms to [13]. The LR04 benthic $\delta^{18}O$ stack of Lisiecki and Raymo [14] is also plotted to underline glacial versus interglacial periods. Label of the marine isotopic stages after [14].

In the central Arctic basin, high carbonate contents have generally been interpreted to reflect interglacial conditions: their record is limited to discrete occurrences in carbonate rich layers correlated to a higher influence of warm Atlantic Ocean waters [from 15 to 19]. On the contrary, barren samples represent glacial periods when there is little to no deposition or preservation of calcareous material [13, 20, 21] (Figure 1).

Another important characteristic of planktonic foraminifera arctic populations is that they present a very low species diversity, most commonly with assemblages being strongly dominated by the sinistral form of the species *Neogloboquadrina pachyderma* (e.g., from [22 to 24], Figure 2). Other species, as *Globigerina bulloides* and *Globigerina* (i.e. *Turborotalita*) *quinqueloba*, are also encountered in the assemblages [e.g., 10], but their recovery is poor in most Arctic sediments [12, 13, 22] (for *G. quinqueloba*, analyses have furthermore to be conducted in the silt-sized fraction of the sediment, i.e. smaller than 150 μm, to record a reliable index of presence of this species [7]). They could however constitute additional clues of northward penetration of warm water masses in the Arctic, especially during interglacials [12, 25, 26].

The recovery of monospecific assemblages can be limiting for the reconstruction of paleoenvironments. However, the species *N. pachyderma* offers great possibilities, when considering the various morphological variants [e.g., 22, 27, 28] which could carry specific paleoenvironmental signatures [i.e., 29, 30].

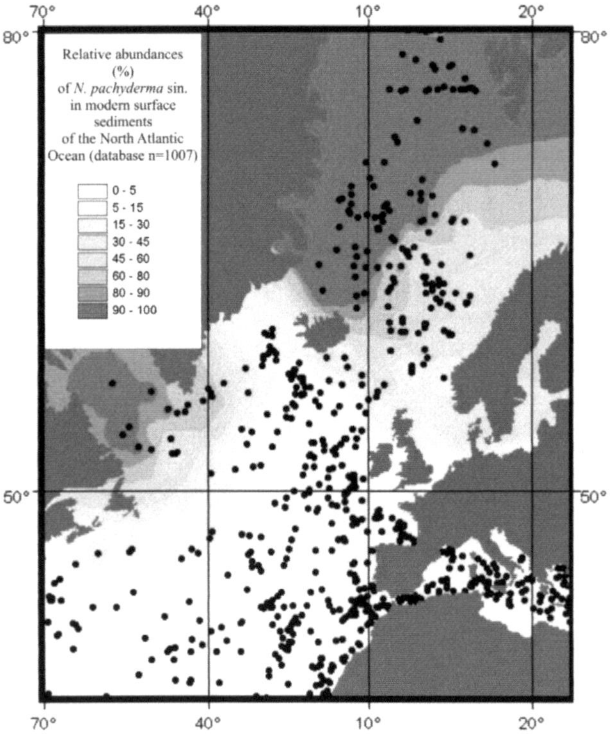

Figure 2. Distribution of the polar foraminifera *N. pachyderma* sinistral in modern surface sediments of the North Atlantic Ocean. Dots mark the geographic position of the modern sediment samples (database from MARGO project, compilation of planktonic foraminifera census data after [31, 32], n=1007 surface sediment samples). Relative abundances were mapped with ARCVIEW.

3. Modern populations of *N. pachyderma*

The species *N. pachyderma* is the most characteristic high-latitude taxon [e.g. 23, 27], comprising more than 90% of the recent assemblages from the Polar Regions of both hemispheres (Figure 2). Some studies have even demonstrated its ability to live in sea ice (as observed in Antarctic sea-ice, [e.g., 33]). Until recently, palaeontological approaches identified two coiling directions in the species *N. pachyderma*, each having a distinct biogeographical range, at least in the North Atlantic Ocean. The change in coiling direction occurs around a mean summer sea-surface temperature (SST) (July-

August-September) of 11°C, a mean winter SST (January-February-March) of 6°C, and a mean annual SST of 8°C (Figure 3). North of this limit, mainly sinistral specimens of *N. pachyderma* are observed (Figure 2). This is one of the key characteristics used in early paleoenvironmental studies [34], as well as recent ones [e.g. 35] to identify major hydrological change through time.

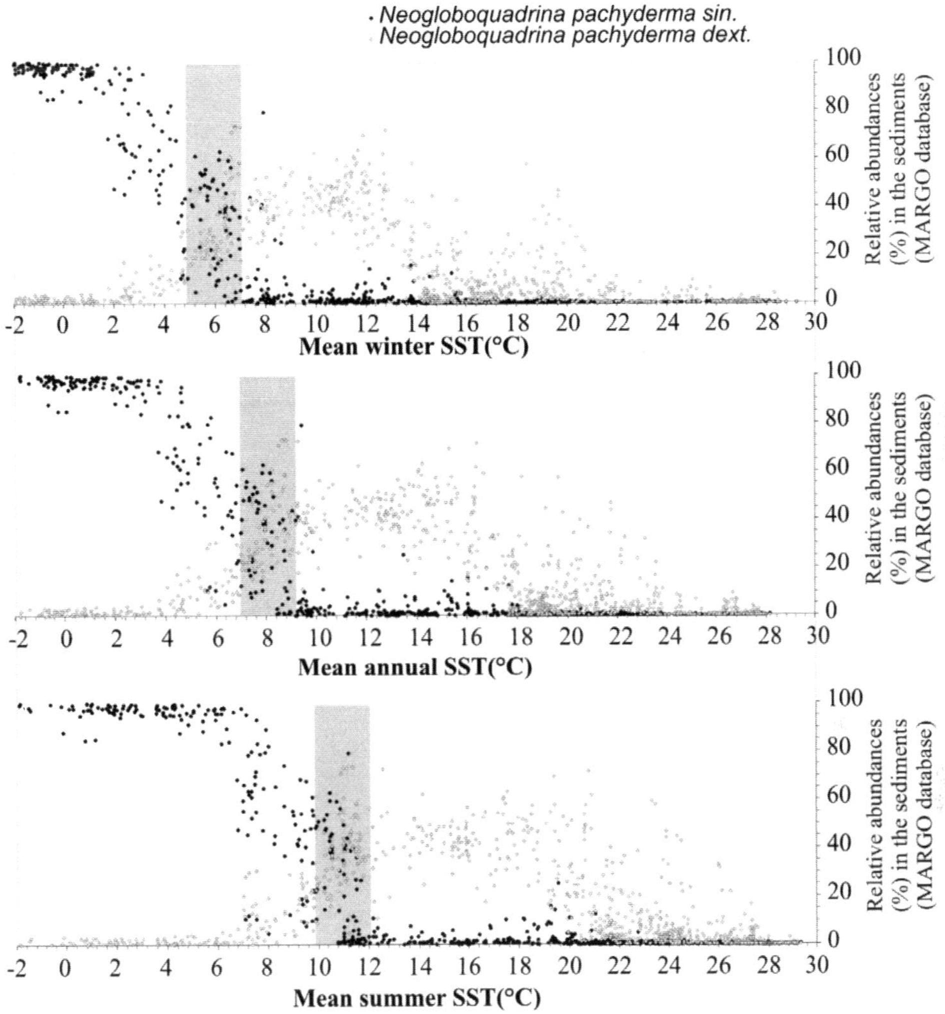

Figure 3. Distribution of *N. pachyderma* sin. versus *N. pachyderma* dex. -syn.: *N. incompta*, (sensu Darling et al., 2006) as a function of sea-surface temperature (SST) in the North Atlantic Ocean (database from MARGO [36], compilation of planktonic foraminifera census data after [31, 32], n=1007 surface sediment samples). Modern hydrological parameters (SST) requested from WOA [37] using the tool developed by Schaffer-Neth during the MARGO project (http://www.geo.uni-bremen.de/geomod/Sonst/Staff/csn/woasample.html).

N. pachyderma sin. taxonomy has recently been revisited throughout molecular approaches [e.g. 23], which have demonstrated the existence of different small subunit ribosomal (SSU) genotypes within the micropaleontogical definition of the *N. pachyderma* morphospecies. According to Darling et al. [23], the dextral and sinistral varieties represent two distinct and divergent lineages, with *N. pachyderma* limited to the dominantly sinistral lineage (the dextral form of *N. pachyderma* should be renamed *N. incompta*). Cryptic genetic types exist within *N. pachyderma* sin. [24, 38], showing that Arctic and Antarctic *N. pachyderma* sin. populations have been isolated throughout the Quaternary [38]. A higher diversity is recorded in the Antarctic, the Arctic domain harbouring only one species [24].

The morphospecies *N. pachyderma* shows a remarkably high degree of morphological variability. This has been known for a long time [e.g., 22, 27, 28, 39], but it has never been systematically investigated by micropaleontologists, and the taxonomic and/or ecological significance of the morphological variants therefore remain unclear [see 29]. A first step toward the consideration of morphotypes was done with the geochemical investigation of different size classes (150-250 μm; >250 μm) of monospecific *N. pachyderma* samples from the western Arctic Ocean (Chukchi Sea, [40]). This study revealed that size selection of specimens permits to unambiguously identify different isotopic signatures. Hillaire-Marcel et al. [40] suggested that this difference was linked to distinctive depth habitats (ecological niches) of tiny versus large specimens of *N. pachyderma* sin. within the water column. They demonstrated that a reverse relationship between specimen weight, or size, and $\delta^{18}O$ values could be observed in the modern Chukchi Sea, with large specimens depicting an offset towards lighter $\delta^{18}O$ values. This signature was attributed to warmer and deeper temperature habitats, with large specimens preferentially calcifying in underlying waters originating from the North Atlantic (a 2.5°C temperature increase is observed from 100 m down to 250 m along the pycnocline in the western Arctic).

Depth related morphological changes were suggested very early for *N. pachyderma* sin. arctic populations [22]. Especially, crystalline thickening of the test, with evidence for an encrustation done organically, was evidenced for these populations on the basis of the comparison of plankton towns versus sediment samples [22, 39]. Complementary, a mesopelagic affinity was thus deduced for these populations on the basis of oxygen isotopic analyses (in relation to their calcification depth: [41, 42]; see [8] for a review). Recent studies demonstrated the importance of the seasonal stratification of the water column on the vertical distribution of *N. pachyderma* sin. modern populations ([30] conducted in the central Irminger Sea).

4. *N. pachyderma* central Arctic morphotypes

Analyses of Central Arctic sediments (IODP-ACEX Hole 4C, [43]) have revealed the existence and the preservation in the sediments of at least five *N. pachyderma* sin. different morphotypes. A template for the classification of these morphotypes was recently and tentatively furnished by Eynaud et al. [29]. Five morphotypes have thus been distinguished based on morphological investigation of specimens under SEM and light microscope (Plate 1 to 4). Criteria allowing their distinction are summarized below.

N. pachyderma sin. morphotypes are characterized as follows: (1) *Nps*-1: classical small-sized specimen with 4 tiny compact chambers (Plate 2; Image 1 to 3); (2) *Nps*-2: 4 globular chambers, prominent apertural lip and large aperture, square shaped shell (Plate 2; Image 6 to 8); (3) *Nps*-3: 4 to 4.5 globular chambers, large aperture with or without lip, elongate shells (Plate 3; Image 1 to 3); Nps-4: 4.5 to 5 globular chambers, prominent apertural lip and large aperture, large shell size (Plate 3; Image 6 to 8); *Nps*-5: 4 chambers, deeply incised sutures, losange shaped shell, aberrant sinistral *N. incompta*? (Plate 4; Image 1 to 3). The wall texture and the degree of encrustation [e.g., 8] is also a discriminating feature of the different morphotypes (Plate 2 to 4, see also [29]).

The morphological diversity within the sinistral variety was first attributed to the ontogeny, assuming that larger specimens represented more mature individuals. On the other hand, it has been demonstrated that small specimens with a reduced last chamber had achieved full adult maturity [6, 44]. Several hypotheses can be further considered, including stress linked both to horizontal or vertical gradients in the water masses. In the Southern Ocean, a graded scheme is indeed observed with a distribution of morphotypes corresponding to oceanic frontal structures that occur from the Sub-Antarctic to the Antarctic domain [27]. These observations are in this domain consistent with those provided by molecular approaches [23]). We can infer that a similar situation applies to the peri- to central-Arctic Ocean domains, with possibly the existence of distinct genotypes. This has not yet been tested by molecular approaches, but however only one genotype has currently been recognized in peri-arctic regions [e.g., 24] despite the occurrence of multiple morphotypes.

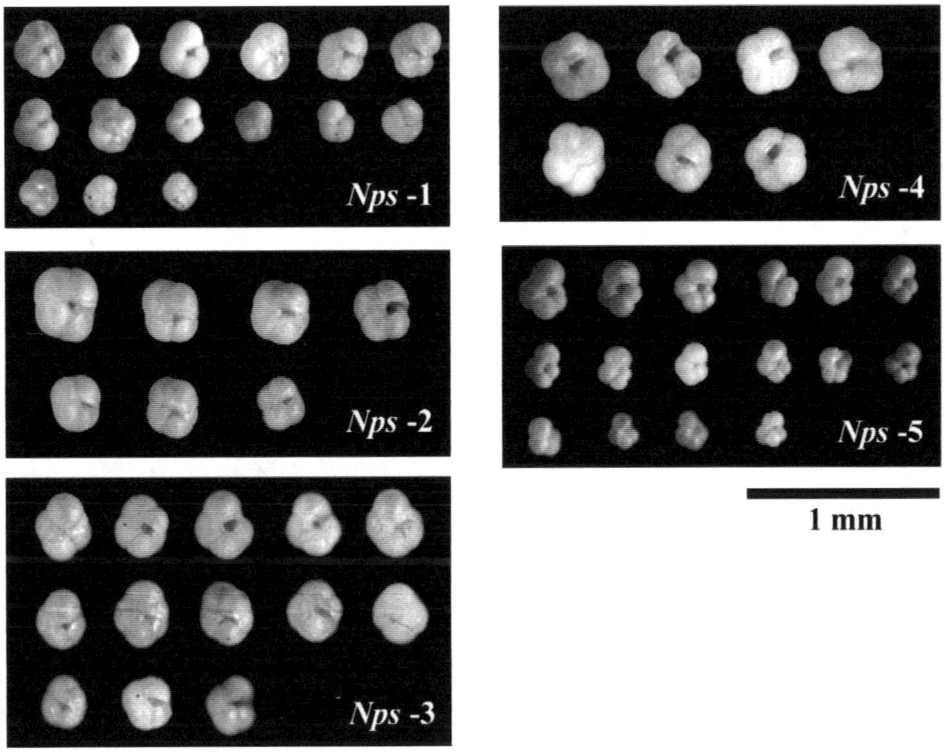

Plate 1. IODP ACEX 302- *Neogloboquadrina pachyderma* sinistral (*Nps*) morphotypes: ventral views from *Nps*-1 to 5. Light Microscope imagery (LEICA Multi-z DM6000).

The most significant feature of Arctic *N. pachyderma* sin. specimens is probably the observation of large types morphologically very similar to subarctic and subtropical types from the same Neogloboquadrinid clade. Interestingly, the timing of the adaptation of N. pachyderma sin. to the Arctic was discussed by Huber al. [45], based on the observation of size enlargement of specimens during the last 0.4 Ma. It is probable that the morphological diversity in the Arctic *N. pachyderma* sin. population is linked to pulsed Atlantic glacial/interglacial inflow, with large specimens calcifying during increased rates of sub-surface penetration of the Atlantic waters as previously documented by Hillaire-Marcel et al. [40] and de Vernal et al. [42].

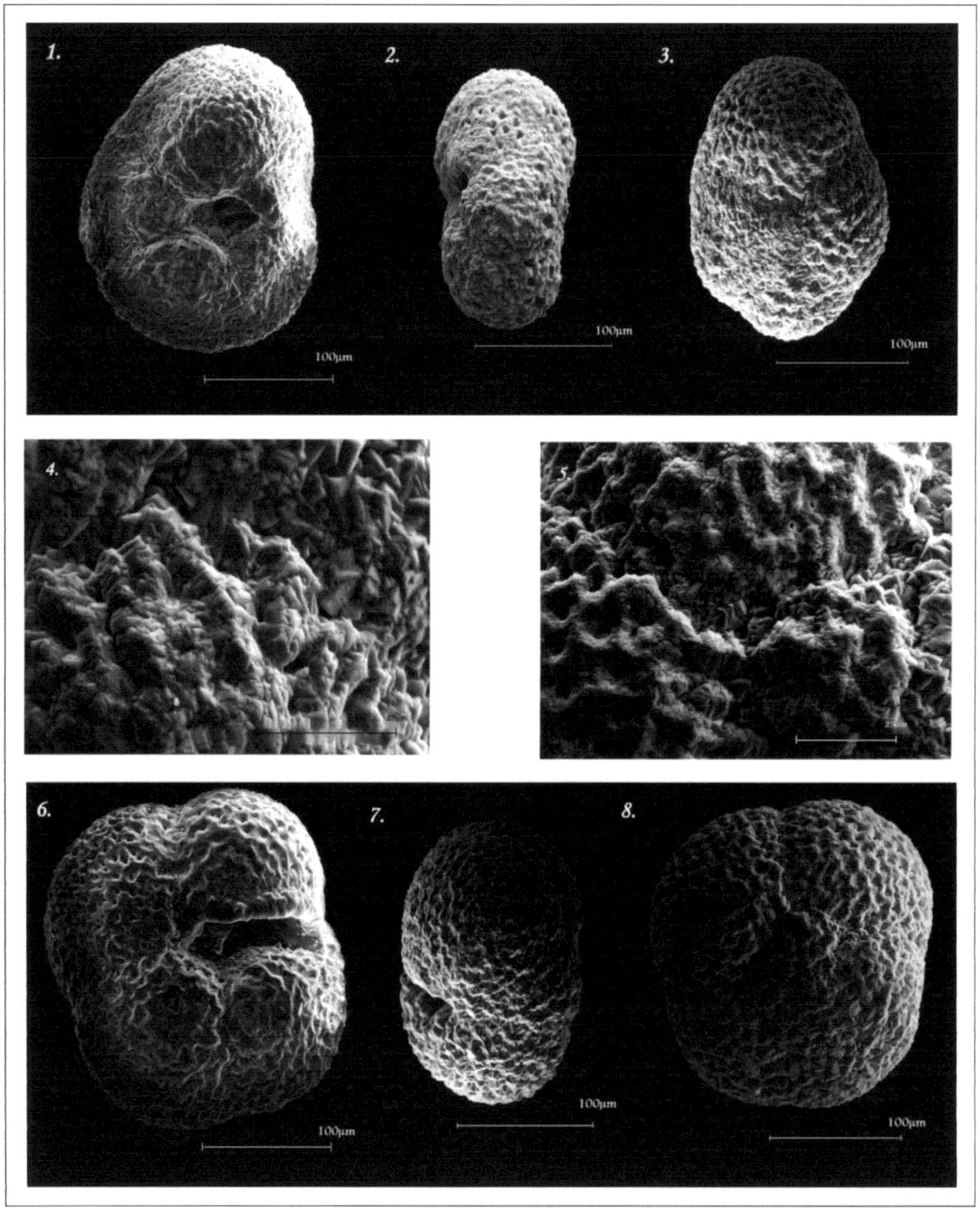

Plate 2 Scanning electron photomicrographs from *Neogloboquadrina pachyderma* sin morphotypes 1 and 2, observed in Arctic sediments; Image 1: ventral view from *Np*s 1; Image 2: edge view from *Np*s 1; Image 3: dorsal view from *Np*s 1; Image 4: close-up view of the test wall structure of *Np*s 1; Image 5: close-up view of the test wall structure of *Np*s 2; Image 6: ventral view from *Np*s 2; Image 7: edge view from *Np*s 2; Image 8: dorsal view from *Np*s 2 Photography credits: GEOTOP [46]

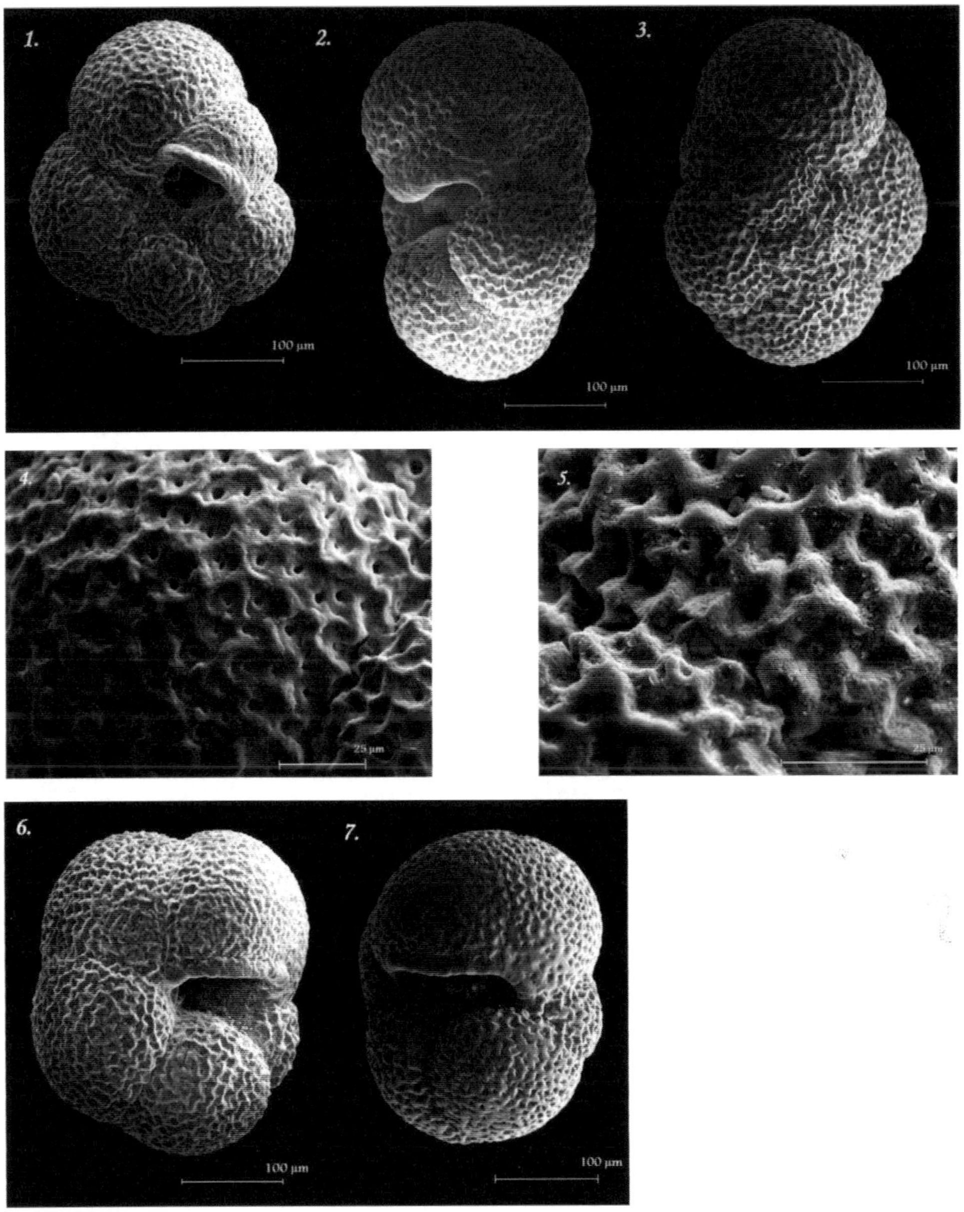

Plate 3 Scanning electron photomicrographs from *Neogloboquadrina pachyderma* sin morphotypes 3 and 4, observed in Arctic sediments; Image 1: ventral view from *Np*s 3; Image 2: edge view from *Np*s 3; Image 3: dorsal view from *Np*s 3; Image 4: close-up view of the test wall structure of *Np*s 3; Image 5: close-up view of the test wall structure of Np-s 4; Image 6: ventral view from *Np*s 4; Image 7: edge view from *Np*s 4 Photography credits: GEOTOP [46]

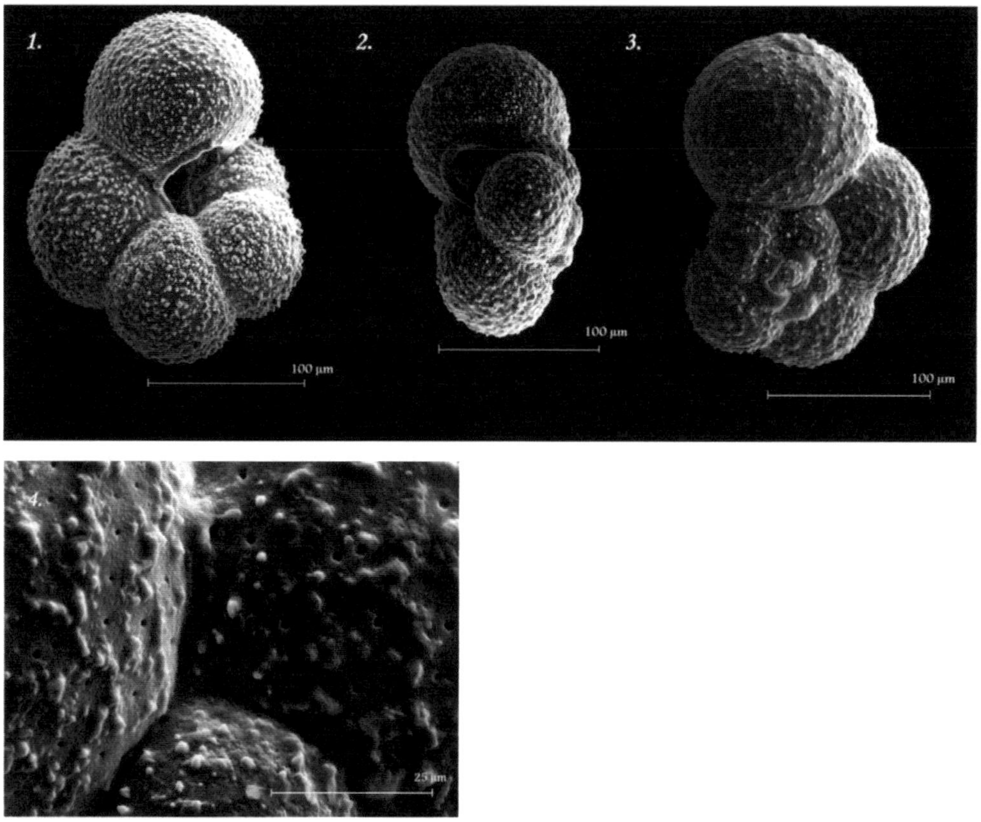

Plate 4 Scanning electron photomicrographs from *Neogloboquadrina pachyderma* sin morphotype 5, observed in Arctic sediments; Image 1: ventral view from *Nps* 5; Image 2: edge view from *Nps* 5; Image 3: dorsal view from *Nps* 5; Image 4: close-up view of the test wall structure of *Nps* 5; Photography credits: GEOTOP [46]

5. Conclusion

In spite of the large dominance of the single species *N. pachyderma* sin. in sediments of the Arctic Ocean, something which could be limiting for paleoceanographic reconstructions, a possibility may exist to qualitatively document past hydrological variability of this extreme environment on the basis of this species. Actually, detailed morphometric investigations, with the discrimination of morphotypes within *N. pachyderma* sin. fossil populations, could provide an alternative way to reconstruct past oceanic circulation. Coupled with geochemical analyses, this investigation could potentially permit to test the efficiency of the exchange of this basin with the North Atlantic Ocean.

During decades, micropaleontologists have tried to summarize the morphological differences among specimens to constitute coherent groups and species. Molecular approaches have recently shown that morphological similarities could mask genetically divergent populations. In the Arctic, the use of the species *N. pachyderma* sin., which dominates the modern and past interglacial assemblages for the last 1.8 Ma, may force us to consider minor morphological criteria to extract precious (paleo)ceanographic information and thus better understand the climatic evolution of this domain.

6. Acknowledgments

This paper synthesizes results obtained in the frame of the Integrated Ocean Drilling Program (IODP) for the Arctic Coring Expedition (ACEX LEG 302). Special thanks are due to Olivier Ther and Robert Escobedo for technical assistance, and to Lucie Barré and the GEOTOP laboratory for SEM imagery (Plate 2 to 4). This is an UMR EPOC contribution N° 1819.

References

[1] Kucera M 2007 Planktonic Foraminifera as Tracers of Past Oceanic Environments *Proxies in Late Cenozoic Paleoceanography (Developments in Marine Geology* vol 1*)* ed C Hillaire-Marcel and A de Vernal (Amsterdam: Elsevier) chapter 6 pp 213–62

[2] Intergovernmental Panel on Climate Change (IPCC) 2007 Climate Change 2007: The Physical Science Basis *Contribution of Working Group I to the Fourth Assessment Report of the Intergovernmental Panel on Climate Change* ed S Solomon *et al.* (Cambridge: Cambridge University Press) 996 pp

[3] Stroeve J, Holland MM, Meier W, Scambos T and Serreze M 2007 Arctic sea ice decline: Faster than forecast *Geophys. Res. Lett.* **34** L09501

[4] Wang M and Overland JE 2009 A sea ice free summer Arctic within 30 years? *Geophys. Res. Lett.* **36** L07502

[5] Serreze MC and JA Francis 2006 The Arctic amplification debate *Clim. Change* **76** 241–64

[6] Hemleben C, Spindler M and Anderson O R 1989 *Modern Planktonic Foraminifera* (New York: Springer) 363 pp

[7] Carstens J and Wefer G 1992 Recent distribution of planktonic foraminifera in the Nansen Basin Arctic Ocean *Deep-Sea Res.* **39** S507–S524

[8] Kohfeld K E, Fairbanks R G, Smith S L and Walsh I D 1996 *Neogloboquadrina pachyderma* (sinistral coiling) as paleoceanographic tracers in polar oceans: Evidence from Northeast Water Polynya plankton tows sediment traps and surface sediments *Paleoceanography* **11** 679–99

[9] Bauch D, Carstens J and Wefer G 1997 Oxygen isotope composition of living *Neogloboquadrina pachyderma* (sin) in the Arctic Ocean *Earth Planet. Sci. Lett.* **146** 47-58

[10] Carstens J, Hebbeln D and Wefer G 1997 Distribution of planktic foraminifera at the ice margin in the Arctic (Fram Strait) *Marine Micropal.* **29** 257–69

[11] Clark DL 1971 Arctic Ocean Ice Cover and Its Late Cenozoic History *Geol. Soc. Am. Bull.* **82** 3313–24

[12] Adler RE, Polyak L, Ortiz JD, Kaufman DS, Channell JET, Xuan C, Grottoli AG, Sellen E and Crawford KA 2009 Sediment record from the western Arctic Ocean with an improved Late Quaternary age resolution: HOTRAX core HLY0503-8JPC Mendeleev Ridge *Global Planet. Change* **68** 18–29

[13] Cronin TM, Smith S, Eynaud F, O'Regan M and King J 2008 Quaternary Paleoceanography of the Central Arctic based on IODP ACEX 302 Foraminiferal Assemblages *Paleoceanography* **23** PA1S18

[14] Lisiecki L E and M E Raymo 2005 A Pliocene- Pleistocene stack of 57 globally distributed benthic 18O records *Paleoceanography* **20** PA1003

[15] Herman Y 1969 Arctic Ocean Qaternary microfauna and its relation to paleoclimatology *Palaeogeogr. Palaeoclimatol. Palaeecol.* **6** 251–76

[16] Herman Y and Hopins D M 1980 Arctic Oceanic climate in late Cenozoic time *Science* **209** 557–62

[17] Spiegler D 1996 Planktonic foraminifer Cenozoic biostratigraphy of the Arctic Ocean Fram Strait (Sites 908-909) Yermak Plateau (Sites 910-912) and East Greenland Margin (Site 913) *Proc. ODP (Ocean Drilling Program) Sci. Results* vol 151, ed J Thiede A M Myhre J V

Firth G L Johnson and W F Ruddiman (College Station TX: Ocean Drilling Program) pp153–67

[18] Huber R, Meggers H, Baumann K-H and R Henrich 2000 Recent and Pleistocene carbonate dissolution of the Norwegian-Greenland Sea *Mar. Geol.* **165** 123–36

[19] Polyak L, Bischof J, Ortiz J, Darby D, Channell J, Xuan C, Kaufman D, Lovlie R, Schneider D and Adler R 2009 Late Quaternary stratigraphy and sedimentation patterns in the western Arctic Ocean *Global Planet. Change* **68** 5–17

[20] Jakobsson M, Løvlie R, Arnold E, Backman J, Polyak L, Knutsen J and Musatov E 2001 Pleistocene stratigraphy and paleoenvironmental variation from Lomonosov Ridge sediments central Arctic Ocean *Global Planet. Change* **31** 1–22

[21] Spielhagen R F, Baumann K-H, Erlenkeuser H, Nowaczyk N R, Nørgaard-Pedersen N, Vogt C and Weiel D 2004 Arctic Ocean deep-sea record of northern Eurasian ice sheet history *Quat. Sci. Rev.* **23** 1455–83

[22] Bé AWH 1960 Some observations on Arctic planktonic foraminifera *Cushman Found. Foram. Research Contr.* **11** 64–8

[23] Darling K F, Kucera M, Kroon D and Wade C M 2006 A resolution for the coiling direction paradox in Neogloboquadrina pachyderma *Paleoceanography* **21** PA2011

[24] Darling K F, Kucera M and Wade C M 2007 Global molecular phylogeography reveals persistent Arctic circumpolar isolation in a marine planktonic protist *Proc. Natl Acad.Sci. USA* **104** 5002–07

[25] Bauch HA, Erlenkeuser H, Fahl K, Spielhagen R F, Weinelt M S, Andruleit H and Henrich R 1999 Evidence for a steeper Eemian than Holocene sea surface temperature gradient between Arctic and sub-Arctic regions *Palaeogeogr. Palaeoclimatol. Palaeecol.* **145** 95–117

[26] Hald M, Ebbesen H, Forwick M, Godtliebsen F, Khomenko L, Korsun S, Olsen LR and Vorren TO 2004 Holocene paleoceanography and glacial history of the West Spitsbergen area Euro-Arctic margin *Quat. Sci. Rev.* **23** 2075–88

[27] Kennett J P 1968 Latitudinal variation in *Globigerina pachyderma* (Ehrenberg) in surface sediments of the southwest Pacific Ocean *Micropaleontology* **14** 305–18

[28] Cifelli R 1973 Observations on *Globigerina pachyderma* (Ehrenberg) and *Globigerina incompta* Cifelli from the North Atlantic *J. Foraminiferal Res.* **3** 157–66

[29] Eynaud F, Cronin T M, Smith S, Zaragosi S, Mavel J, Mary Y, Mas V and Pujol C 2009 Morphological variability of the planktonic foraminifer *Neogloboquadrina pachyderma* from ACEX cores: Implications for Late Pleistocene circulation in the Arctic Ocean *Micropaleontology* **55** 101–16

[30] Jonkers L, Brummer G-J A, Peeters F J C, van Aken H M and De Jong M F 2010 Seasonal stratification shell flux and oxygen isotope dynamics of left-coiling *N pachyderma* and *T quinqueloba* in the western subpolar North Atlantic *Paleoceanography* **25** PA2204

[31] Kucera M, Weinelt M, Kiefer T, Pflaumann U, Hayes A, Weinelt M, Chen M-T, Mix AC, Barrows TT, Cortijo E, Duprat J, Juggins S and Waelbroeck C 2005 Reconstruction of sea-surface temperatures from assemblages of planktonic foraminifera: Multi-technique approach based on geographically constrained calibration data sets and its application to glacial Atlantic and Pacific Oceans *Quat. Sci. Rev.* **24** 951–98

[32] Hayes A, Kucera M, Kallel N, Sbaffi L and Rohling EJ 2005 Glacial Mediterranean sea surface temperatures based on planktonic foraminifera assemblages *Quat. Sci.e Rev.* **24** 999-1016

[33] Spindler M 1996 On the salinity tolerance of the planktonic foraminifer *Neogloboquadrina pachyderma* from the antarcic sea-ice *Proc. NIPR Symp. Polar Biol.* **9** 85–91

[34] Ericson D B 1959 Coiling direction of *Globigerina pachyderma* as a climatic index *Science* **130** 219–20

[35] Eynaud F, de Abreu L, Voelker A, Schönfeld J, Salgueiro E, Turon J-L, Penaud A, Toucanne S, Naughton F, Sanchez-Goñi M-F, Malaizé B an Cacho I 2009 Position of the Polar Front

along the western Iberian margin during key cold episodes of the last 45 ka *Geochem. Geophys. Geosyst.* **10** Q07U05

[36] MARGO (Multiproxy Approach for the Reconstruction of the Glacial Ocean surface) project members 2009 Constraints on the magnitude and patterns of ocean cooling at the Last Glacial Maximum *Nature Geoscience* doi:101038/ngeo411

[37] WOA 1998 World Ocean Atlas 1998 National Oceanographic Data Center Silver Spring Maryland http://wwwnodcnoaagov/oc5/woa98html

[38] Darling K F, Kucera M, Pudsey C J and Wade C M 2004 Molecular evidence links cryptic diversification in polar plankton to Quaternary climate dynamics *Proc. Natl Acad. Sci. USA* **101** 7657–62

[39] Vilks G 1973 A study of *Globorotalia pachyderma* (Erhenberg) = *Globigerina pachyderma* (Erhenberg) in the Canadian Arctic *PhD Dissertation* Department of Oceanography Dalhousie University Halifax Nova Scotia

[40] Hillaire-Marcel C, de Vernal A, Polyak L and Darby D 2004 Size-dependent isotopic composition of planktic foraminifers from Chukchi Sea vsNW Atlantic sediments - implications for the Holocene paleoceanography of the western Arctic *Quat. Science Rev.* **23** 245–60

[41] Simstich J, Sarnthein M and Erlenkeuser H 2003 Paired $\delta^{18}O$ signals of *Neogloboquadrina pachyderma* (s) and *Turborotalita quinqueloba* show thermal stratification structure in Nordic Seas *Marine Micropal.* **48** 107–25

[42] de Vernal A, Hillaire-Marcel C and Darby D A 2005 Variability of sea ice cover in the Chukchi Sea (western Arctic Ocean) during the Holocene *Paleoceanography* **20** PA4018

[43] Moran K *et al.* 2006 The Cenozoic palaeoenvironment of the Arctic Ocean *Nature* **441** 601–05

[44] Vilks G 1975 Comparison of Globorotalia pachyderma (Ehrenberg) in the water column and sediments of the Canadian Arctic *J. Foraminiferal Res.* **5** 313–25

[45] Huber R, Meggers H, Baumann K H, Raymo M E and R Henrich 2000 Shell size variation of the planktic foraminifer *Neogloboquadrina pachyderma* sin in the Norwegian-Greenland Sea during the last 13 Ma - implications for paleoceanographic reconstructions *Palaeogeogr. Palaeoclimatol. Palaeecol.* **160** 193–212

[46] Barré L 2009 Composition (13C et 18O) des morphotypes de Neogloboquadrina pachyderma dans les sédiments de l'océan Arctique *Master Thesis* Université Bordeaux 1

Marine diatoms in polar and sub-polar environments and their application to Late Pleistocene paleoclimate reconstruction

Xavier Crosta

UMR-CNRS 5805 EPOC, Université Bordeaux 1, Avenue des Facultés, 33405 Talence Cedex, France

E-mail: x.crosta@epoc.u-bordeaux1.fr

Abstract. Diatoms are one of the major phytoplankton groups in polar and sub-polar marine environments along with green algae and chrysophytes. Diatoms are composed of two components, a two-valve test made of amorphous silica and an organic cell encapsulated into the test. Mucilage covering the test and proteins embedded in the silica lattice of the test completes the organic pool of the diatoms. The preservation of these two components into deep-sea sediments allows for a large set of diatom-based proxies to infer past oceanographic and climatic changes in polar and sub-polar marine environments.

Most diatom species in polar and sub-polar marine environments exhibit a narrow range of ecological preferences, especially in terms of sea-surface temperature and sea ice conditions. Preserved diatom assemblages in deep-sea sediments mirror the diatom assemblages in the phytoplankton. It is subsequently possible to extrapolate the relationships between diatom assemblages in surface sediments and modern parameters to down-core fossil assemblages to document past changes in sea-surface temperatures and sea ice conditions. Congruent analysis of biogenic silica and organic carbon and stable isotope ratios (O, Si in the silica matrix and C, N in the diatom-intrinsic organic matter) provides information on siliceous productivity, nutrient cycling and water mass circulation. Measurements of diatom biomarkers give complementary information on sea ice conditions and siliceous productivity.

1. The diatoms

1.1. Classification of diatoms

Diatom comes from the Greek *diatomos*, which means ‘cut in two’ or ‘cut in half’.

Diatoms are unicellular organisms in which the cell is encapsulated in an amorphous silica box, called the frustule, composed of two intricate valves (the larger one is called the epivalve and the smaller one the hypovalve). Diatom size varies from 2 μm to 1-2 mm. Diatoms belong to the class Bacillariophyceae, in which two orders were defined based on the shape of the frustule: Centrales (round diatoms) and Pennales (elongated diatoms). The Order Centrales is divided in three suborders, each characterized by the shape of the cells, the polarity and the arrangement of the processes. The Order Pennales is divided in two suborders, one including diatoms with a raphe (an elongated slit or pair of slits through the valve wall), the other one including diatoms without a raphe [1].

The frustule is highly ornamented with processes (labiate, strutted, internal or external, with or without extensions), pores (areolae), costae, spines, horns, setae and other distinguishing features. Diatom taxonomy is historically based on the morphological characteristics and ornamentation of the frustule [2]. Diatom taxonomy is in constant re-evaluation since the emergence of electronic microscopes and genetic techniques. It is worth noting that paleoceanographers using diatoms are limited to differentiation based on the light microscope.

1.2. Biology of diatoms

Diatoms mostly reproduce through vegetative division at a rate of 0.1-8 times per day [3]. Vegetative reproduction involves the formation of two new hypovalves in the parent diatom's frustule, which progressively leads to reduce the average size of the diatom population. At a given threshold, physiologically dictated, diatoms undergo sexual reproduction through gamete fusion and the formation of an auxospore that renews a full-sized vegetative cell [4]. Size diminution continues in diatoms that did not reproduce sexually until they eventually die.

Some species have another particular stage, the resting spore, which is formed under unfavorable conditions (depleted nutrient levels, low light levels, etc...) and allows the diatom to survive until better conditions return. In the polar marine environment, resting spores allow the diatom population to survive the polar winter and are an important inoculum for spring/summer blooms [5].

1.3. Ecology of diatoms

About 285 genera and 12000 species of diatoms have been identified [2] although these numbers are in constant evolution. Diatoms are found in almost every aquatic environment, including fresh and marine waters. Diatoms are also found in aerophilic environments, such as moss banks. Diatoms are photosynthetic organisms possessing yellow-brown chloroplasts with pigments including chlorophyll a and c, β-carotene, fucoxanthin, diatoxanthin and diadinoxanthin that enable them to capture a wide range of wavelengths and some species are able to live at very low light levels [6]. Most diatoms are non-motile though raphid species use mucilage secretion along their raphe for motility. Diatoms are restricted to the photic zone. In the open marine environment, diatoms are generally planktonic, although some benthic or pseudo-benthic species attached to macro-algae or sea-ice are also encountered.

In the world ocean, diatoms mainly thrive in cold, nutrient-rich regions where silicic acid is not limiting, such as polar regions as well as upwelling systems and coastal areas. In other regions, diatoms are outcompeted by carbonate and organic organisms that have lower nutrient requirements.

At the regional scale, many factors interact to determine the distribution of planktonic diatoms. In polar and sub-polar areas, the most important factors are sea-surface temperatures, sea-ice conditions, macro- and micronutrient levels, stability of the surface water layer, light levels, and grazing. Salinity may also exert a major role on diatom distribution, especially in coastal regions of the Artic Ocean where there is a strong freshwater input from major rivers.

1.4. From production to burial

In polar oceans, 1-10% of the diatoms produced in surface waters reach the sediment [7]. This ratio is higher in shallow coastal areas than in abyssal open ocean regions. Biotic factors such as sedimentation type (single particles vs aggregates or fecal pellets, mass sedimentation events) and abiotic factors such as lateral transport and dissolution in the water column and at the water-sediment interface are the main processes determining the diatom flux to the sea-floor. These processes not only favor the preservation of highly silicified diatoms but also alter the geochemical signals embedded into the diatoms. It has however been shown that the residual sedimentary assemblages are still indicative of surface conditions in different oceanic regions such as the North Pacific [8], the Southern Ocean [9-11] and the northern North Atlantic [12]. Diatoms can therefore be used to infer past oceanographic and climatic changes in these regions.

2. Diatoms as paleoceanogrpahic tools

Due to the specificity of diatom morphology (organic matter attached and embedded into the siliceous frustule) and ecology, diatoms can be used in various ways to infer past oceanographic conditions including methods based on micropaleontology (diatom identification), geochemistry (biogenic content, organic biomarkers) and isotope geochemistry (either in the diatom-intrinsic organic matter or the siliceous matrix).

2.1. Micropaleontology

2.1.1. *Rationale.*In polar oceans, diatom distribution in phytoplankton presents strong zonations [13-15] and diatom distribution in surface sediments presents similar zonations [16-18]. Processes responsible for the production and zonation of diatoms in phytoplankton and export and preservation of diatoms in marine sediments are briefly described in the previous paragraphs. In summary, it is possible to draw direct relationships between diatom assemblages in surface sediments and modern parameters, such as for example sea-surface temperatures (SST) and sea ice duration/concentration (SIP). It is then possible to extrapolate the relationships observed between diatom assemblages in surface sediments and modern parameters to down-core fossil assemblages in order to document past changes in oceanography, in siliceous productivity and, ultimately, in climate. Though our understanding of diatom ecology is based on investigations of diatom distribution in phytoplankton and sediment traps, it is important to note that paleoceanographic reconstructions are based on altered sedimentary diatom assemblages. It is therefore essential to document and quantify the relationships that exist between diatom distribution in surface sediments and environmental parameters as a preliminary step to any paleoceanographic study.

The hypothesis linking diatom assemblages in fossil down-core sediments and surface parameters is called actualism. A given diatom assemblage is produced, and preserved, under specific modern conditions. If the same assemblage is found down-core, then oceanographic and climatic conditions may have been the same in the past as they are today. As such, the use of diatom assemblages to reconstruct past climate variability is more straightforward for recent times when modern species were extant than for older periods when species that are extinct today were an important component of the diatom assemblages.

Diatom identification has been heavily used in polar regions to reconstruct several parameters over the last 40-50 million years: sea-surface temperature [19, 20], sea ice cover [21, 22], siliceous productivity [23, 24], ocean circulation [25] and relative sea level [26]. Diatom identification is also widely used for biostratigraphy [27, 28]. Investigations can be based on down-core variations of given species of known ecology or on statistical methods that approximate the relationships between modern species and their ecological preferences. As such, transfer functions can provide quantitative estimates of surface properties (SST and sea ice conditions in polar regions) that are essential to constrain or verify paleoclimatic models.

A transfer function must be understood as any kind of mathematical approach that analyses census counts of fossil assemblages to produce absolute values of surface properties by comparing fossil samples to a subset of modern samples having definite modern conditions. Transfer functions can work on reduced species data sets, but transfer functions applied to marine diatom from polar environments generally work with 20-40 species. Diatom-based transfer functions in the marine polar realm are mainly dedicated to the reconstruction of SST and SIP though it is also possible to estimate sea-surface salinity [29] and sea level [26]. It is worth noting that it is technically possible to infer any variable that has a statistically significant relationship to diatom species distribution.The most common ones are the Imbrie and Kipp Method (IKM) [30], the Modern Analog Technique (MAT) [31], the Weighted Averaging Partial Least Square (WA-PLS) [32], Maximum Likelihood (ML) [33] and the Artificial Neural Network method (ANN) [34, 35]. The General Additive Model (GAM) [36] and the Revised Analog Method (RAM) [37] are variations of the IKM and MAT approaches, respectively.

Transfer functions work in three steps. First, a calibration data set, consisting of counts of diatom taxa in modern sediments and instrumentally recorded values of environmental parameters at the corresponding locations, is assembled. Second, a mathematical model is developed to quantify the relationship between the environmental variable (parameters) and the census counts. Third, this mathematical model is applied to fossil diatom counts to produce a quantitative estimate of the environmental parameter.

Some points to take into account when developing a transfer function:

- Ubiquitous species presenting no relationship with environmental parameters may not be included in the modern database as they will add noise to the estimate.
- Species that do not account for more than 2% of the diatom assemblage in one sample of the modern sediments are generally not included in the modern database though there is no theoretical reason to exclude them. However, the mathematics supporting function transfer would not benefit from this additional information as the statistical weights of rare species will be overwhelmed by dominant species.
- Dominant species strongly affect transfer functions that use factor analysis as a starting point. For example, *Fragilariopsis kerguelensis* accounts for > 60% of the diatom assemblage in Southern Ocean surface sediments when SST is between 2-7°C. As a result, IKM approach estimates around 4°C over the whole range of highest *F. kerguelensis* relative abundances in the training set (validation of the transfer function to estimate the distribution field of modern parameters). The overwhelming influence of dominant species can be alleviated through normalisation of diatom relative abundances by class ranking [38], logarithmic transformation [39], square root or Hellinger. Dominant species do not impact comparative approaches such as the MAT, for which rare species with low abundances are as equally important as dominant species.
- Mathematical techniques must be used according to the calibration data set. Reduced data sets in terms of number of samples and geographical extent would work better with techniques using factor analysis that calculate a mathematical function between the core-top samples and the parameters, thus coping with the lack of samples. Conversely, large calibration data sets work better with more descriptive techniques such as the MAT.
- Paleoreconstructions might be different (1) for a given mathematical technique depending on the calibration data set and (2) for a given calibration data set depending on the mathematical technique.

More details about transfer function techniques, validation, advantages and disadvantages are presented in [40] and [41].

2.1.2. *Application.*Reconstruction of past physical parameters (SST and SIP) in the polar oceans is of prime importance for a sound understanding of Earth's internal climate system. Quantification of SST and SIP in the past is also essential to constrain or validate paleoclimatic models. In the Southern Ocean, diatoms are certainly the best proxy to reconstruct these parameters while in the Northern Hemisphere dinocysts and foraminifera can be equally used. As such diatoms can be used alone through investigations of single species abundances or transfer functions that nicely compliment qualitative studies. It is worth noting that, although some discrepancies may exist between diatom-based paleoreconstructions of SST and SIP as noted above, much greater differences are generally apparent between diatom-based paleoreconstructions and other microfossil-based transfer functions or with geochemical approaches. I describe below two examples to illustrate these points.

Coupling qualitative and quantitative approaches for sea ice reconstruction. Sea ice is one of the fastest reacting, most seasonal geophysical parameters on the Earth's surface of which the waxing and waning strongly influence the global climate. This seasonal cycle strongly affects high latitude ecosystems, economies and civilizations. It also affects Earth's climate through the transfer of high latitude anomalies to lower latitudes via the atmosphere and the ocean. Over the last several decades, there has been a dramatic reduction in perennial sea ice in the Arctic Ocean [42] and the

Bellinghausen and Amundsen seas in West Antarctica [43]. It is therefore of prime importance to estimate sea ice dynamics in the past and to map sea ice distribution at key periods such as the Last Glacial Maximum (LGM), the present and the last interglacials. To date, the most extensive dataset for Antarctic sea ice mapping was produced for the LGM in the framework of MARGO project (Multiproxy Approach for the Reconstruction of the Glacial Ocean surface). This study represents also a perfect example of coupling of qualitative and quantitative data.

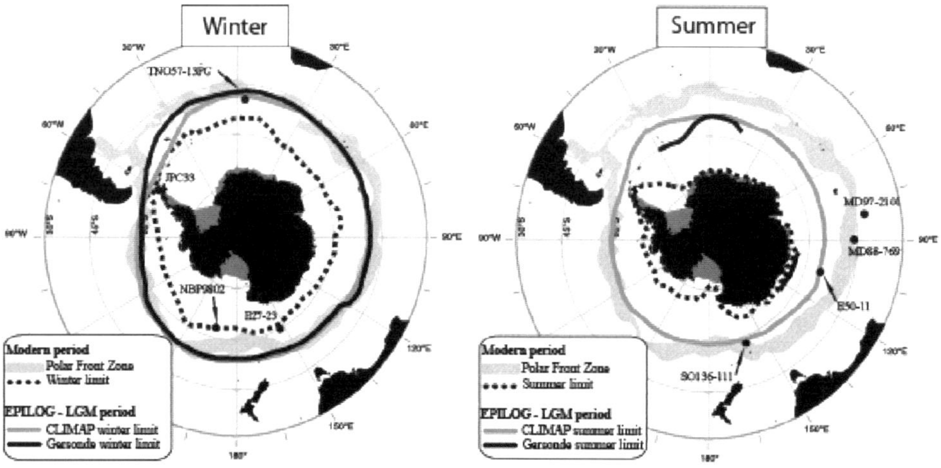

Figure 1. Sea-ice extent at the Last Glacial Maximum during winter and summer seasons in the context of the MARGO program (modified from [45]). Gersonde's winter sea-ice limit was calculated from the *F. curta* group proxy [44] and the MAT approach [21], while Gersonde's summer limit was estimated from the *F. obliquecostata* proxy [44] and the MAT approach [21]. CLIMAP's winter and summer limits were located at the faunally identified winter and summer 0°C isotherms, respectively [46], and lithogenic tracers such as changes in the sedimentation rate, presence of ice-rafted debris and geographical contact between diatom oozes and clayed sediments [50]. Position of the modern sea-ice limits is from Schweitzer [51] and location of the modern hydrographic fronts is from Orsi et al. [52].

Based on extensive investigations of time-series sediment-traps and diatom distribution in surface sediment of the Weddell Sea, Gersonde and Zielinski [44] showed that relative abundances of the *Fragilariopsis curta* group (*F. curta* and *F. cylindrus*) greater than 3% of the total diatom assemblage indicated the presence of winter sea ice. They also showed that relative amounts of *Fragilariopsis obliquecostata* greater than 3% indicated the presence of summer sea ice. A combination of the single species proxies described above with a diatom-based transfer function approach (31 species used, Modern Analog Technique [21]) provided a comprehensive picture of winter and summer sea ice cover at the LGM [45]. Diatom assemblages demonstrated that winter sea ice cover at the LGM was twice the modern winter surface while the summer sea ice surface at the LGM was 2-3 times the modern summer surface (Figure 1). These findings greatly refined estimates from CLIMAP [46] in the early 1980s, especially about summer sea ice extent. Indeed, CLIMAP mapped the summer sea ice limit at around 55°S in the Atlantic and Indian sectors of the Southern Ocean. But it was demonstrated shortly thereafter that CLIMAP estimated spring sea ice limit rather than summer sea ice limit [47]. When CLIMAP sea ice maps are used as boundary conditions, paleoclimatic models estimate a 80 ppm drop in atmospheric CO_2 concentrations [48], while when diatom-based sea ice reconstructions are used, paleoclimatic models estimate a 10-30 ppm drop in atmospheric CO_2 concentrations due to sea ice insulation [49].

Quantitative approaches for sea-surface temperature reconstructions. Here again, diatom-based transfer functions for SST reconstructions were first developed for the Southern Ocean [38], but rapidly exported for the northern North Atlantic [53]. Diatom-based transfer functions in the Southern Ocean are mainly based on IKM and MAT, while a suite of diatom-based transfer functions have been developed for the North Atlantic. It is worth noting that, there, the new techniques such as WA-PLS and ML did not produce drastically different estimates than IKM [54], either in terms of absolute values or patterns. However, standard errors or Root mean square errors of prediction (RMSEP) were smaller with the new mathematical approaches.

As opposed to the Southern Ocean where diatoms are the main microfossil group for paleoclimatic reconstructions, foraminifera and dinocysts can be used in the northern North Atlantic to estimate SST and SIP. However, diatom-based SST estimates have been produced in many cores from the northern North Atlantic. Diatom-based SST estimates are sometimes rather different than dinocyst-based SST from the same core [55, 56], possibly due to their different ecological response or seasonal production. Most diatom-based SST records in the northern North Atlantic are restricted to the Holocene Epoch because of the scarcity of diatoms older than the last deglaciation.

The Holocene is the baseline for recent warming. Understanding the regional heterogeneity of climate changes under different forcings, in a well dated framework, may help to understand the modern and future ocean-atmosphere response. Signal-to-noise rations are smaller during the Holocene Epoch compared to glacial-interglacial cycles and it is necessary to gain insight on how perform transfer functions.

Hereafter is an example on how perform several diatom-based transfer functions applied to a similar down-core diatom dataset. IKM, WA-PLS and ML transfer functions have been applied to diatom assemblages in a sediment core from the Norwegian Sea that covers the last 13 ka BP [54]. The three transfer functions documented a similar SST trend over the last 13 ka BP. Some discrepancies were however apparent between IKM and the two other techniques in terms of the amplitude of changes. IKM estimated warmer summer SST during the Early Holocene and somehow colder summer SST during the Late Holocene with a sharp drop at around 5-4 ka BP. A similar sharp drop in SST was observed in the Southern Ocean at ~4 ka BP [57, 58]. WA-PLS and IKM both use several components with an additional step for WA-PLS that maximizes the covariance between the environmental variables to be reconstructed and, hence, the predictive power of the method in contrast to the I&K method where the components are chosen irrespective of their predictive value to capture the maximum variance within the biological data. The ML transfer function method has no parameters to set, and has the least autocorrelation problems. This transfer function method can be chosen if many of the taxa show a unimodal relationship with the environmental variable of interest [59]. The present example demonstrated that SST estimates are significant, almost irrespective of the transfer function used, when the modern datasets are extended enough to adequately cover the full range of environmental conditions expected in the past.

In the present example, the three reconstructions documented an overall cooling trend (Figure 2), in step with the decreasing insolation in the Northern Hemisphere, indicating a causal impact of orbital forcing on Holocene sea-surface temperatures in the Nordic seas [55]. A focus on the 13-9 ka BP period documented sea-surface temperatures 7-9°C colder during the Younger Dryas than today and a rapid warming at 11.7 ka BP in phase with the warming recorded in Greenland ice cores. The slow warming registered after the sharp transition between the Younger Dryas and the Pre-Boreal Optimum was interrupted by two cooling events attributed to disturbances in the North Atlantic Ocean thermohaline circulation. Superimposed on the general trend, high-amplitude centennial variability was observed in the sea surface temperatures (Figure 1). The variability showed a periodicity of 80-120 years, similar to solar activity periodicities indicating that even small changes in solar irradiance can have a large impact on sea-surface temperatures and ocean circulation [20]. Similar periodicity was identified in relative abundances of sea ice-related diatoms in a sediment core from the Antarctic continental shelf and was similarly attributed to solar activity forcing [58].

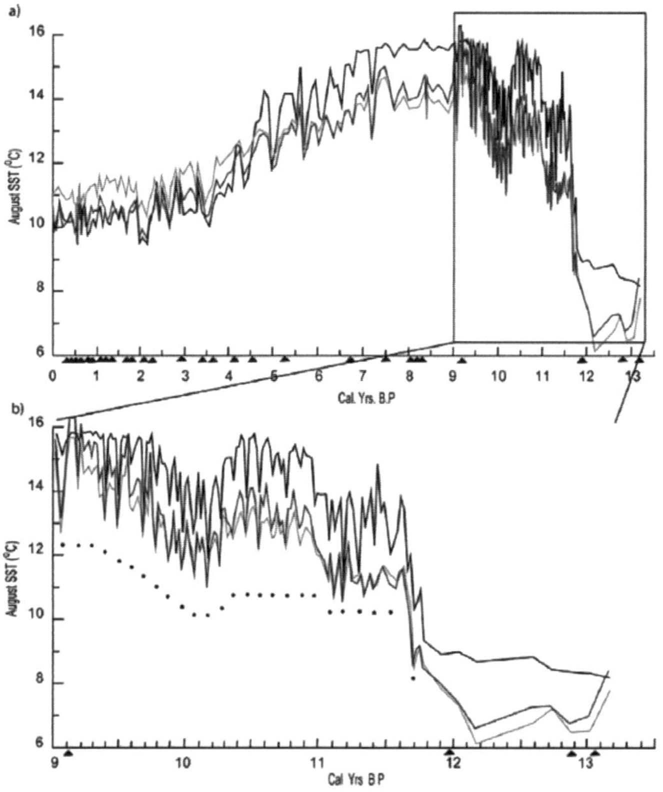

Figure 2. Diatom-based sea surface temperature reconstructions during the last 13 ka (Imbrie and Kipp, black curve; Weighted Averaging Partial Least Square, red curve; Maximum Likelihood, green curve). The period from 13-9 ka BP is shown in details in the lower graph. The black dots denote significant cooling events, while the black triangles represent ^{14}C dating Reproduced with permission from [54] (© Berner et al., 2010), by permission of SAGE Publications Ltd.

Quantitative approaches for other parameters. Diatom-based transfer functions are more rarely used for sea-surface salinity [29] and relative sea level changes [26] in the northern North Atlantic. This may in part reflect difficulties in developping such transfer functions because of the low signal-to-noise ratio in coastal areas. Indeed, diatom assemblages in coastal areas generally show strong variability, sometime irrespective of salinity and sea level, because of other parameters that are more important in these highly dynamic zones (wind mixing, lateral transport, tides, etc…). However, these transfer functions, if further improved in the future, will give important quantitative data to paleoclimatic models.

2.2. Geochemistry

2.2.1. Rationale. Diatoms transfer biogenic silica and organic matter, including species-specific molecules, to deep sea sediments. When dissolution/preservation issues are constrained, it is possible to use geochemical proxies to reconstruct past changes in environmental conditions. The content of biogenic silica and organic carbon in sediments tracks variations in productivity of siliceous organisms [60], which in turn gives information on the oceanic nutrient cycles [61]. Trace element content

(Zn/Si; Ge/Si) records changes in surface water salinity and delivery rates of weathering products to the sea from continental and seafloor weathering, respectively [61, 62]. Finally, content in diatom-specific biomarkers (highly branched isoprenoids; long-chain diols), which are synthesized by very few diatom species living in specific habitats, reflect changes in sea ice cover and productivity, respectively [63, 64].

In conclusion, diatoms offer a suite of geochemical proxies to compliment diatom census counts and diatom-based transfer functions. Combining these different approaches is essential in regions where other organisms are not well preserved in sediments. Additionally, radiocarbon dating of diatom bound organic compounds allows the replacement of paleoceanographic interpretations from carbonate-barren sediments back into a temporal timescale [65].

2.2.2. Applications

Qualitative reconstructions of sea ice conditions. Recently, it has been shown that some diatoms present in Arctic sea ice biosynthesize a specific biomarker, a mono-unsaturated highly branched isoprenoid (HBI) composed of 25 atoms of carbon [63]. A di-unsaturated HBI (Diene) is also synthesized by Arctic sea ice diatoms while isomers with greater number of unsaturation (Triene to Pentaene) are synthesized by phytoplankton diatoms. However it has been shown that the monoeme isomer was the most stable in sediments underneath Arctic sea ice and it was, thus, proposed as a new proxy to infer Arctic spring sea ice. This proxy is referred as IP25 [63]. As a concentration proxy, one has to be careful about dilution of the monoeme by other sedimentary phases (biogenic organisms or terrigenous particles), especially in coastal areas that are prone to material advection and winnowing. Laboratory methods, implying extraction with a solvent, purification of the extract and analysis by GC-MS, are described in Belt et al. [63, 66]. Analyzing HBI content in down-core sediments is much faster than identifying diatom assemblages. However HBIs fail to capture of the environmental details that can be drawn from the identification of dozens of diatom species or species groups.

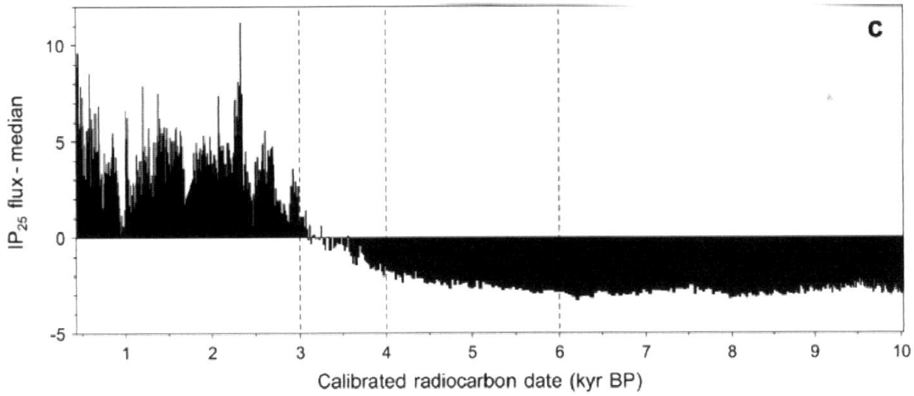

Figure 3. Temporal profile of biomarkers in the ARC-3 core (Canadian Arctic Archipelago) versus time, presented as the difference between IP25 fluxes and the median. Reprinted from [68] with permission from Elsevier.

Several IP25 records have been published in the Arctic realm to date [67, 68]. One of these records from the Canadian Arctic Archipelago demonstrated that spring sea ice was lowest during the early-to-mid Holocene, increased slowly between 6-4 ka BP and increased abruptly between 4-3 ka BP to reach a modern, though highly variable sea ice occurrence, at around 2.5 ka BP (Figure 3) [68].

Diatom HBIs evidenced a similar evolution of the sea-ice cover in the Antarctic realm during the Holocene [69, 70], in phase with diatom census counts from the same cores. In the Southern Ocean, sea ice diatoms synthesize a di-unsaturated HBI (Diene) while phytoplankton diatoms synthesize a triunsaturated HBI (Triene). Environmental interpretations are based on the Diene-to-Triene ratio [71]. It has been suggested that early diagenesis does not affect differently the Diene and Triene, though more work at the seasonal timescale is needed to really determine how the HBI signal is formed and exported from the photic zone and preserved in modern and down-core sediments. The Diene-to-Triene ratio is additionally free of any dilution effect. As such, it represents a complimentary and quick proxy to document Antarctic sea ice conditions, interpreted as qualitative yearly sea ice duration. Here again, diatom census counts offer a wider range of paleoenvironmental interpretations.

The transition between a warmer Hypsithermal and a cooler Neoglacial at around 5-4 ka BP has been evidenced globally by other proxies [72], possibly in relation to the disappearance of northern Hemisphere ice sheets [73] and ocean reoganisation [74].

Productivity over the last 30 ka. Diatoms play a major role in the export of organic carbon to the deep sea. Studying the factors controlling the contribution of diatoms to total production is pivotal for our understanding of the mechanisms controlling the efficiency of the biological pump of carbon in various regions of today's ocean. Only 1-10% of the biogenic silica produced in surface waters is eventually preserved in deep-sea sediments. Grazing, dissolution, remineralization, export efficiency and diagenesis strongly alter the surface signal. Nonetheless, it was shown that oceanic zones with high silicic acid content and high siliceous productivity such as the Southern Ocean, the North Pacific and upwelling systems display high biogenic silica in surface sediments [7]. As such, it is believed that down-core biogenic silica content (and more accurately, down-core biogenic silica fluxes) can be used to track past changes in siliceous productivity. Comparison of biogenic content to organic carbon content in deep sea cores opens a new field on diatom physiology changes through time [75, 76]. Comparison of biogenic silica content to stable isotopes measured in diatoms give insight in past nutrient utilization and past oceanic circulation [77, 78].

Diatom biomarkers are an emerging analytic and scientific field to track changes in diatom productivity. Long-chain diols (C28 and C30), biomarkers specifically synthesized by the diatom genus *Proboscia*, may be used to estimate past changes in nutrient input by Southern Ocean upwelling [64].

The two examples below highlight how diatom-based organic geochemistry can give information on productivity changes during the last 30 kans. Obviously, the biogenic silica proxy has been applied to much older sediments to represent siliceous productivity changes over the geological time scale in relation to major oceanographic reorganizations [79, 80]. Examples were here chosen to fit the Late Pleistocene glacial-interglacial timescale.

The Southern Ocean is one of the largest High-Nutrient, Low-Chlorophyll (HNLC) regions of the world in which low concentrations of trace metals, such as iron, limit primary productivity. The North Pacific is also a HNLC region, though iron input is greater than in the Southern Ocean because the latter one has less proximal continents [81]. Today, high productivity (mainly diatoms) is found south of 50°S between the Antarctic Polar Front and the winter sea ice edge due to the upwelling of deep waters that brings nitrate and silicic acid to the surface waters [82]. High productivity is also encountered in Antarctic coastal areas where iron limitation is alleviated [83].

Analysis of opal content in three cores from the three sectors of the Antarctic Ocean demonstrated a large increase in opal fluxes during the deglaciation (Figure 4). A concomitant increase in $^{231}Pa/^{230}Th$ ratios indicate that these greater opal fluxes did not result from changes in preservation but reflect changes in opal production as a result of a more intense upwelling [23]. Both ^{231}Pa and ^{230}Th are produced uniformly by radioactive decay of dissolved uranium. Both isotopes are removed from seawater by scavenging. However, ^{231}Pa has a strong affinity to sorption to biogenic opal particles while ^{230}Th has a more even affinity for sorption to any sinking particle. As such, $^{231}Pa/^{230Th}$ ratios of particulate material throughout the global ocean are positively correlated with the opal content of particles. Particulate $^{231}Pa/^{230}Th$ ratios are not impacted by loss of biogenic phases during diagenesis. Downcore records of particulate $^{231}Pa/^{230Th}$ can thus be interpreted in terms of past changes in opal fluxes. In the present example, the positive correlation between $^{231}Pa/^{230Th}$ and opal fluxes indicates that deglacial maximum in opal flux reflected changes in opal production and silicic acid supply rather than variable opal preservation.

However, the balance between greater opal and organic carbon production (drawdown) and greater upwelling (outgassing) resulted in greater atmospheric CO_2 concentration demonstrating a direct link between ventilation of the deep Southern Ocean and atmospheric CO_2 concentrations [84].

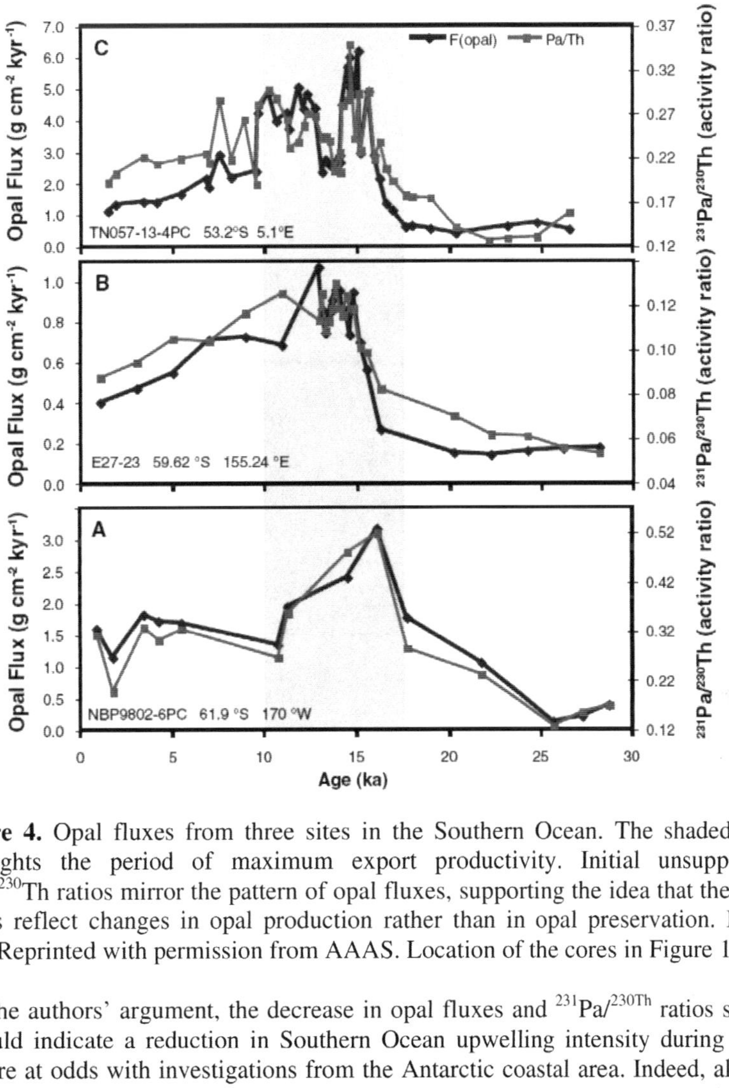

Figure 4. Opal fluxes from three sites in the Southern Ocean. The shaded box highlights the period of maximum export productivity. Initial unsupported $^{231}Pa/^{230}Th$ ratios mirror the pattern of opal fluxes, supporting the idea that the opal fluxes reflect changes in opal production rather than in opal preservation. From [23]. Reprinted with permission from AAAS. Location of the cores in Figure 1.

Following the authors' argument, the decrease in opal fluxes and $^{231}Pa/^{230Th}$ ratios since 10 ka BP (Figure 4) would indicate a reduction in Southern Ocean upwelling intensity during the Holocene. These results are at odds with investigations from the Antarctic coastal area. Indeed, although overall productivity decreased during the Holocene in Antarctic coastal areas (Figure 5a,b), long-chain diols and diatom counts demonstrated enhanced intrusion of Modified Circumpolar Deep Waters to coastal areas (Figure 5c). Long-chain diols are synthesized by *Proboscia* species that thrive in turbulent, nutrient-rich surface waters as evidenced both in Antarctica [64] and the Arabian Sea [85]. These studies indicated that long-chain diols can be used as proxies for high nutrient conditions. Therefore the diol index record from the Antarctic Peninsula showing a strong increase at around 4 ka BP (Figure 5c) suggest a congruent increase in turbulence and nutrient input to surface water and, thus, an increase in upwelling of Modified Circumpolar Deep Waters. Similar increase in upwelling of

Modified Circumpolar Deep Waters was suggested in East Antarctic coastal area based on sedimentological and diatom census counts proxies [24].

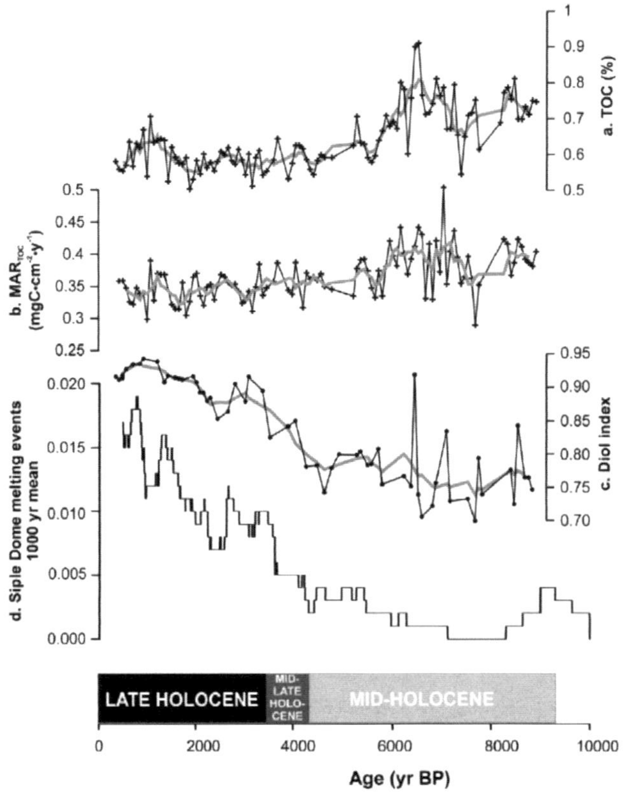

Figure 5. Antarctic climate records of **a.** TOC content of JPC-33 core, and **b.** Mass accumulation rates of TOC (MARTOC) reflecting general productivity. **c.** Long-chain diol index of JPC-33 core reflecting *Proboscia* diatom productivity. **d.** Melt frequency observed at Siple Dome ice core (after Das & Alley 2008). Dark grey lines in **a**, **b**, and **c** represent running average with a 10% smoothing window (position of the core in Figure 1). From [64].

Two possible explanations may reconcile the apparent paradox of lower upwelling in southern hemisphere mid- and high latitudes. First, the core of Southern Ocean upwelling moved southward during the Holocene in phase with the poleward migration of the Westerlies [86]. As a result, less nutrients were brought to the Polar Front Zone area. Second, paleoproductivity (understood as export and burial of organic carbon and silica) was decoupled from nutrient input and resulted from greater export of larger diatoms. It is worth noting that $\delta^{15}N_{diat}$ and $\delta^{30}Si$ records do not show any special feature (a peak demonstrating greater nutrient utilization or a low suggesting greater nutrient input) during the deglaciation when the peak in opal flux is recorded [76, 77, 78, 87, 88].

2.3. Isotope geochemistry

2.3.1. Rationale. It is possible to analyze stable isotopic ratios of light elements in diatoms to track changes in surface water properties. Several different and complementary isotopes can be measured in diatoms. Two groups of isotopes detected in diatoms can be differentiated: firstly, oxygen (O) and silicon (Si) isotopes that are carried by the diatom frustule and, secondly, carbon (C) and nitrogen (N) isotopes that are carried by the organic matrix. This organic matrix, called diatom-intrinsic organic matter (DIOM), is intimately embedded in the silica lattice where it directs opal biomineralization. Analyzing DIOM rather than bulk organic matter provides a more direct picture of surface water nutrient cycling because the DIOM is protected from remineralisation and diagenesis by the silica matrix. $\delta^{18}O$ provides information on the water masses, $\delta^{30}Si$ and $\delta^{15}N$ on nutrient cycling (as a balance between supply and uptake) and $\delta^{13}C$ on nutrient cycling and, in given conditions, on CO_2 exchange at the ocean-atmosphere interface.

Measurements of $\delta^{18}O_{diatom}$ reflect the isotopic composition of the ambient seawater, which is a function of changes in global ice volume as well as local changes in evaporation, freshwater inputs, salinity and changes in water mass or ocean circulation, the temperature of the water in which the diatom is precipitated and the presence of any isotope vital effects that may exist in diatoms. Measurements of $\delta^{18}O_{diatom}$ are representative of changes in the photic zone [89]. Measurements of $\delta^{18}O_{diatom}$ were successfully used to track iceberg melting events in the Southern Ocean [90, 91].

The $\delta^{15}N$ of sinking organic matter reflects the utilization of nitrate in regions where nitrate is not completely consumed [92]. Phytoplankton preferentially takes up ^{14}N-bearing NO_3^-. As the initial NO_3^- supply is progressively consumed, the $\delta^{15}N_{nitrate}$ increases, leading to a related increase in the $\delta^{15}N$ of the organic matter produced from the NO_3^-. The N isotopic composition of the surface NO_3^- pool and surface sedimentary organic matter were shown to reflect the proportion of the NO_3^- supply from the subsurface that is consumed in the euphotic zone. Enriched $\delta^{15}N$ values thus indicate enhanced nitrate relative utilisation.

The same principle and interpretation are applicable to silicon stable isotopic ratios in which enriched $\delta^{30}Si$ values reflect enhanced silicic acid relative utilisation in surface waters [87, 93].

2.3.2. Applications

Productivity and nutrient cycling in the Southern Ocean over the last 30 ka

Nitrogen and silicon stable isotopic ratios measured in diatoms track nutrient relative utilization as a balance between supply (mainly upwelling in the Southern Ocean) and uptake (mainly by diatoms in the Southern Ocean). As such, the important changes in biogenic burial evidenced during deglaciation and the inferred increase in upwelling [23] should be seen in the $\delta^{15}N_{diat}$ and $\delta^{30}Si$. However, no $\delta^{15}N_{diat}$ and $\delta^{30}Si$ investigations from the Antarctic part of Southern Ocean show any significant changes during the 10-15 ka BP (e.g., Figure 6). Important changes in the isotope data occurred prior to the peak in opal. Only records from the Subantarctic to Subtropical zones evidenced a possible concomitant peak in $\delta^{15}N_{diat}$ (Figure 6, right box) [94]. As such N and Si isotopes are not very supportive of enhanced productivity and upwelling in the Antarctic realm between 15-10 ka BP.

$\delta^{15}N_{diat}$ and $\delta^{15}N_{bulk}$ records from the Antarctic Ocean demonstrated a Holocene pattern toward lighter values (Figure 6) [25] demonstrating a reduction in relative nutrient utilization, in phase with decreasing productivity but increasing upwelling in coastal areas [65].

Coupling nitrogen and silicon isotopes in the same records may help to draw some interpretations about changes in diatom physiology. Such changes need to be taken into account to better understand the relationships between productivity and nutrient input.

Iron content in surface waters has a strong impact on diatom physiology. Laboratory experiments have shown that uptake ratios of N/Si and C/Si by diatoms are greater when iron is not limiting [95, 96]. Disregarding potential modification of the nitrate-to-silicic acid supply ratio by the upwelling of Circumpolar Deep Water, anti-correlated $\delta^{15}N_{diat}$ and $\delta^{30}Si$ records argue for different relative uptake of nitrate and silicic acid through time, with greater uptake of nitrate during the last glacial and greater uptake of silicic acid during the Holocene (Figure 6). These changes are in agreement with the dust concentration record of Vostok, indicating a potential role of iron deficiency alleviation on diatom physiology during glacial times. However, differential nutrient uptake between warm and cold times was not transferred to the Subtropical Zone as little change in $\delta^{15}N_{diat}$ and $\delta^{30}Si$ is recorded at the SSTF (and out of phase with the climatic periods). It is possible that Antarctic Surface Waters sank and were transported northward as Antarctic Intermediate Waters to ultimately fuel low latitude systems. This hypothesis, called the Silicic Acid Leakage Hypothesis, may explain a ~ 40 ppm decrease in atmospheric CO_2 concentrations [97], which is almost half of the glacial drop.

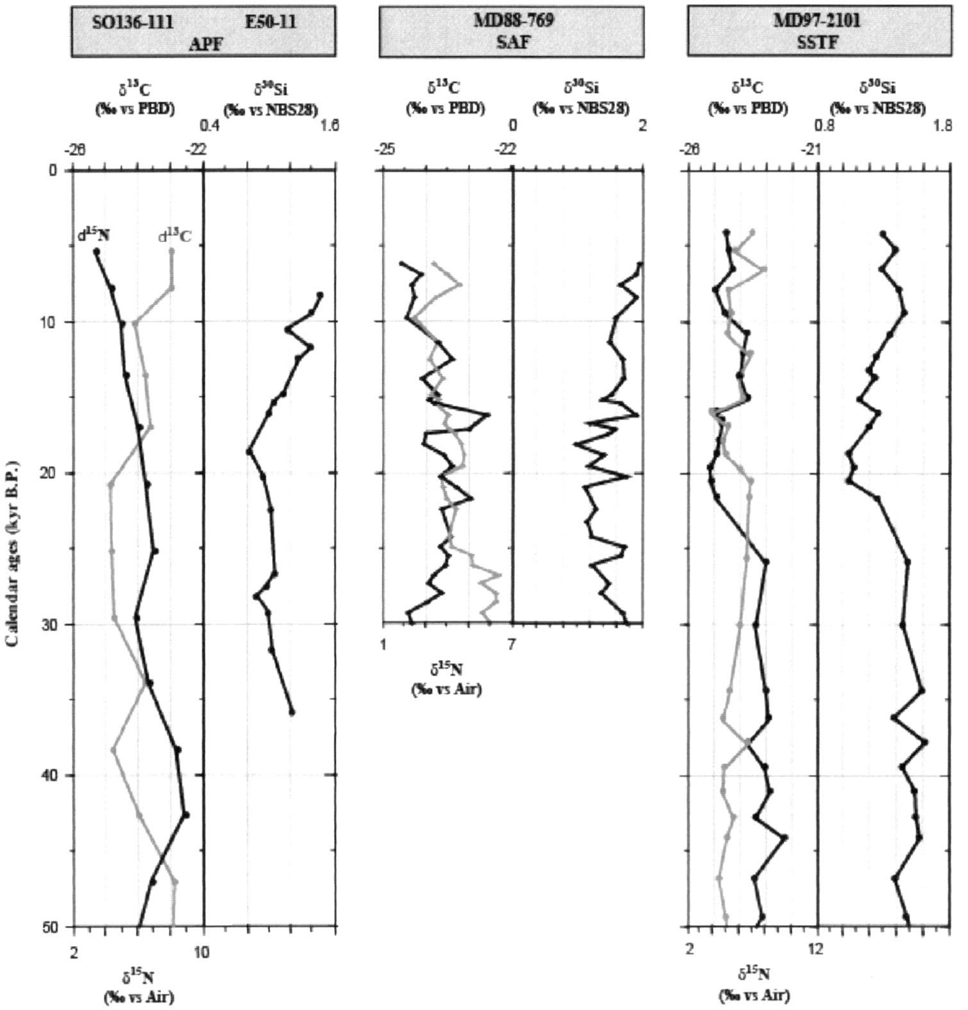

Figure 6. Carbon, nitrogen and silicon isotopes in diatoms versus age from the Antarctic Polar Front Zone (APF), the Subantarctic Front (SAF) and the Southern Subtropical Front (SSTF) in the Indian sector of the Southern Ocean [78, 87, 98]. Reprinted from [41], with permission from Elsevier. Location of the cores in Figure 1.

Productivity and nutrient cycling in the North Pacific over the last 30 ka. Similar to the Antarctic Ocean, large regions of the the subarctic North Pacific are today characterized by high phytoplankton productivity and incomplete consumption of the major nutrients in surface waters. Investigation of deep-sea cores have, however, demonstrated that, during much of the last ice age, biological export production in the subarctic Pacific, including the Bering and Okhotsk Seas, was considerably lower than during the Holocene (Figure 7, upper box) [99]. The pattern of DIOM-$\delta^{15}N$ record is opposite to the opal record with enriched values during the last glacial when opal content was low and depleted values during the Holocene when opal content was high (Figure 7). A constant pool of nutrient over the considered period would have been conducive to depleted $\delta^{15}N_{diat}$ values during the last glacial and enriched $\delta^{15}N_{diat}$ values during the Holocene. The observed anti-correlation between opal content and $\delta^{15}N_{diat}$ over the last 30 ka suggested a reduced supply of nitrate to the surface due to stronger stratification of the upper water column of the Bering Sea during glacial times, possibly in relation to greater sea ice extent such as inferred in the Southern Ocean [100]. The reduction of the nutrient pool during glacial time yielded to a more complete consumption of nitrate and therefore enriched $\delta^{15}N_{diat}$.

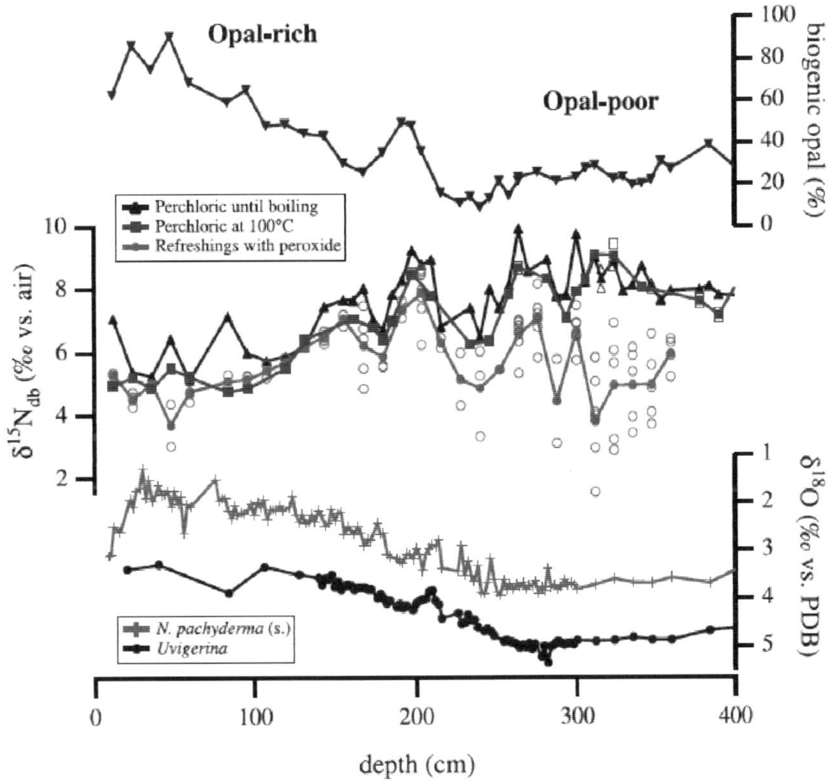

Figure 7. Records of biogenic opal, $\delta^{15}N_{diat}$ and $\delta^{18}O_{foram}$ in a core from the Bering Sea [99]. The last glacial period covers the 400-270 cm interval, the last deglaciation covers the 270-150 cm interval, and the Holocene covers the 150-0 cm interval. Copyright 2007 American Geophysical Union. Reproduced by permission of American Geophysical Union.

3. Summary and future developments of the diatom tool

Diatom identification provides unique environmental reconstructions on sea-surface temperatures, sea ice, salinity and productivity. Diatom census counts have long been used to infer such past oceanic conditions. We are however far from fully understanding this proxy as very few studies monitored diatom production in the phytoplankton, diatom export and preservation in deep-sea sediments over multi-annual time series [101, 102]. It is necessary to better understand diatom ecology when considering shorter spatial and temporal timescales as other factors than large scale parameters may play a significant role on diatom distribution. Multi-annual survey of diatom distribution in phytoplankton and modern sediments is however difficult in polar areas because of logistic issues.

Diatom identification under a light microscope is very time consuming, especially when diatom preservation is poor or when dilution by terrigenous particles is high. As such, one of the next challenges is to develop automated methods for the identification and biometry of diatoms, such as done for palynomorphs [103], coccolithes [104] and foraminifera [105]. An early step was recently done in this direction. An automated imaging and characterization system with computer-controlled transmitted light microscopy adapted to diatom specificities is currently in development [106].

More developments will certainly concern geochemistry. Currently, a great effort is been made to obtain reliable radiocarbon dates from the organic compounds intrinsic to diatom frustules (Ingalls et al., 2004), in order to help in dating sediments devoid of foraminifera. More work is needed to better understand the formation of the isotope signals in diatoms. It is essential to better constrain vital effects that dictate isotopic fractionation factors [107-109], nutrient sources and metabolite pathways [110, 111] and alteration of the signal formed in the phytoplankton during settling and burial [92, 112]. More work is also needed to develop new protocols that allow for accurate and rapid measurement of stable isotopes in diatoms [113-115]. To date, very few records of $\delta^{30}Si$ exist, all of them at multi-millennial scales, partly because of the tedious protocol. Diatom-specific biomarkers will certainly represent the greatest advancement in diatom research in the next decade. However, work is needed to identify precisely which species synthesize HBIs. Mono- and di-unsaturated HBIs are synthesized by some sea ice diatoms such as *Haslea* spp., which represent a minor component of both Arctic and Antarctic sea ice diatom communities [63], while tri- to penta-unsaturated are synthesized by open ocean diatoms such as *Rhizosolenia* spp. Little is known about the role of HBIs in the diatoms. Similarly, long-chain diols are synthesized by *Proboscia* spp. [64], which can represent a significant part of open ocean diatom communities [116]. Here again, little is known about the role of long-chain diols in diatom physiology. It is therefore essential to better understand which species form HBIs or diols, in which conditions, when in the year, and how the molecules are exported and preserved in sediments to refine the interpretations based on these proxies. It is worth noting that the sea ice diatoms synthesizing mono- and di-unsaturated are very seldom found in sediments. With this understanding, the coupling of diatom-biomarkers and diatom census counts will be much more effective and will, maybe, give insight on sea ice seasonality in that sea ice HBIs are thought to be formed during early spring while diatom counts are thought to infer yearly sea ice presence.

References

[1] Simonsen R 1979 *Bacillaria* **2** 9
[2] Round F, Crawford R and Mann D 1990 *The Diatoms* (Cambridge: University Press) 747 p
[3] Smetacek VS 1985 *Mar. Biol.* **84** 239
[4] Round F 1972 *Nova Hedwigia* **23** 291
[5] Horner R 1985 *Sea Ice Biota* (Boca Raton FL: CRC Press) 83
[6] Kemp AES 2000 *Deep-Sea Res.* II **47** 2129
[7] Ragueneau O et al. 2000 *Globa Planet. Change* **26** 317
[8] Sancetta C 1992 *Paleoceanography* **7(2)** 183
[9] Armand L, Crosta X, Romero O and Pichon JJ 2005 *Palaeogeogr. Palaeoclimatol. Palaeoecol.* **223** 93
[10] Crosta X, Romero O, Armand L and Pichon JJ 2005 *Palaeogeogr. Palaeoclimatol. Palaeoecol.* **223** 66
[11] Romero O, Armand L, Crosta X and Pichon JJ 2005 *Palaeogeogr. Palaeoclimatol. Palaeoecol.* **223** 49
[12] Koç-Karpuz N and Schrader H 1990 *Paleoceanography* **5(4)** 557
[13] Hasle GR 1969 *Hvalradets Skrifter* **52** 168 p
[14] Simonsen R 1974 *Meteor Forsch.-Ergeb.* **Reihe D19** 1
[15] Semina HJ 2003 *Iconographia Diatomologica* **10** 1

[16] Defelice DR and Wise SW 1981 *Mar. Micropaleontol.* **6** 29
[17] Sancetta C 1982 *Micropaleontology* **28(3)** 221
[18] Zielinski U and Gersonde R 1997 *Palaeogeo. Palaeoclim. Palaeoecol.* **129** 213
[19] Schneider-Mor A, Yam R, Bianchi C, Kunz-Pirrung M, Gersonde R and Shemesh A 2005 *Geophys. Res. Let.* **32** doi: 10.1029/2005GL022543
[20] Jiang H, Eiricksson J, Schulz M, Knudsen K and Seidenkrantz M 2005 *Geology* **33(1)** 73
[21] Crosta X, Pichon JJ and Burckle L 1998 *Paleoceanography* **13(3)** 284
[22] Justwan A and Koç N 2008 *Mar. Micropaleontol.* **66** 264
[23] Anderson R, Ali S, Bradtmiller L, Nielsen S, Fleisher M, Anderson B and Burckle L 2009 *Science* **323** 1443
[24] Denis D et al. 2009a *Quat. Sci. Rev.* **28(13-14)** 1291
[25] Denis D et al., 2009b *Paleoceanography* **24** doi: 10.1029/2008PA001689
[26] Woodroffe S and Long A 2010 *Quat. Int.* **221(1-2)** 91
[27] Koç N and Scherer R 1996 *Proc. Ocean Drill. Prog., Scient. Results* **151** 61
[28] Zielinski U and Gersonde R 2002 *Mar. Micropaleontol.* **45** 225
[29] De Sève MA 1999 *Mar. Micropaleontol.* **36** 249
[30] Imbrie J and Kipp N G 1971 *The Late Cenozoic Glacial Ages* ed K K Turekian (New Haven: Yale University Press) 71
[31] Hutson WH 1980 *Science* **207** 64
[32] ter Braak CJF and Juggins S1993 *Hydrobiologia* **269-270** 485
[33] Birks CJA and Koç N 2002 *Boreas* **31(4)** 323
[34] Malmgren BA and Nordlund U 1997 *Palaeogeogr. Palaeoclimatol. Palaeoecol.* **136** 359
[35] Malmgren BA, Kucera M, Nyberg J and Waelbroeck C 2001 *Paleoceanography* **16(5)** 520
[36] Armand L 1997 *Ph D Thesis* ANU Canberra 393 p
[37] Waelbroeck C, Labeyrie L, Duplessy JC, Guiot J, Labracherie M, Lelaire H and Duprat J 1998 *Paleoceanography* **13(3)** 272
[38] Pichon JJ, Labeyrie L, Bareille G, Labracherie M, Duprat J and Jouzel J 1992 *Paleoceanography* **7** 289
[39] Zielinski U, Gersonde R, Sieger R and Fütterer D 1998 *Paleoceanography* **13(4)** 365
[40] Kucera M, Rosell-Melé A, Schneider R, Walbroeck C and Weilnet M 2005 *Quat. Sci. Rev.* **24** 813
[41] Crosta X and Koç N 2007 *Proxies in Late Cenozoic Paleoceanography (Developments in Marine Geology* vol 1) ed C Hillaire-Marcel and A de Vernal (Amsterdam: Elsevier) chapter 8 p 327
[42] Polyak L et al. 2010 *Quat. Sci. Rev.* **29** 1757
[43] Adams B et al. 2009 *Antarctic Climate Change and the Environment* SCAR 183
[44] Gersonde R and Zielinski U 2000 *Palaeogeogr. Palaeoclimatol. Palaeoecol.* **162** 263
[45] Gersonde R, Crosta X, Abelmann A and Armand L 2005 *Quat. Sci. Rev.* **24** 869
[46] CLIMAP 1981 *Geol. Soc. Am. Map Chart Ser.* **MC-36**
[47] Burckle L, Robinson D and Cooke D 1982 *Nature* **299** 435
[48] Stephens BB and Keeling RF 2000 *Nature* **404** 171
[49] Morales-Maqueda MA and Rahmstorf S 2002 *Geophys. Res. Lett.* **29(1)** doi: 10.1029/2001GL013240
[50] Cooke D and Hays JD 198. *Antarctic Geoscience* The Union of Wisconsin Press 1017
[51] Schweitzer P 1995 *U.S. Geological Survey Digital Data Series* Virginia
[52] Orsi A, Whitworth T and Nowlin W 1995 *Deep-Sea Res.* **42(5)** 641
[53] Koç N, Jansen E and Haflidason H 1993 *Quat. Sci. Rev.* **12** 115
[54] Berner K, Koç N and Godtliebsen F 2010 *The Holocene* **20(2)** 245
[55] Andersen C, Koç N and Moros M 2004 *Quat. Sci. Rev.* **23** 2155
[56] Solignac S., Giraudeau J and de Vernal A 2006 *Paleoceanography* **21** doi: 10.1029/2005PA001175

[57] Hodell DA, Kanfoush SL, Shemesh A, Crosta X, Charles CD and Guilderson TP 2001 *Quat. Res.* **56** 191
[58] Crosta X, Debret M, Denis D, Courty M-A and Ther O 2007 *Geochem. Geophys. Geosyst.* **8(11)** doi: 10.1029/2007GC001718
[59] Berner K, Koç N, Divine D, Godtliebsen F and Moros M 2008 *Paleoceanography* **23** doi: 10.1029/2006PA001339
[60] Kumar N, Anderson R, Mortlock R, Froelich P, Kubik P, Dittrich-Hannen B and Suter M 1995 *Nature* **378** 675
[61] Shemesh A, Mortlock R and Froelich P 1989 *Paleoceanography* **4(3)** 221
[62] Hendry K, and Rickaby R 2008 *Paleoceanography* **23** doi: 10.1029/2007PA001576
[63] Belt S, Massé G, Rowland S and Rohmer M 2007 *Org. Geochem.* **38** 16
[64] Willmott V, Rampen S, Domack E, Canals M, Sinninghe Damsté J and Schouten S 2009 *Antarctic Science* **22(1)** 3
[65] Ingalls AE, Anderson RF and Pearson A 2004 *Marine Chemistry* **92** 91
[66] Brown TA et al. 2010 *Polar Biology* doi: 10.1007/s00300-010-0942-5
[67] Massé G, Rowland S, Sicre M, Jacob J, Jansen E and Belt S 2008 *Earth Planet. Sci. Lett.* **269** 565
[68] Vare L, Massé G, Gregory T, Smart C and Belt S 2009 *Quat. Sci. Rev.* **28(13-14)** 1354
[69] Barbara L, Crosta X, Massé G and Ther O 2010 *Quat. Sci. Rev.***29** 2731
[70] Denis D, Crosta X, Barbara L, Massé G, Renssen H, Ther O and Giraudeau J 2010 *Quat. Sci. Rev.* **29** 3709
[71] Massé G, Rowland S, Crosta X, Schmidt S and Belt S submitted *Deep-Sea Res. I*
[72] Mayewski PA et al. 2004 *Quat. Res.* **62** 243
[73] Renssen H., Seppä H, Heiri O, Roche D, Goosse H and Fichefet T 2009 *Nat. Geosci.* **2** 411
[74] Debret M, Sebag D, Crosta X, Massei N, Chapron E., Bout-Roumazeilles V and Petit JR 2010 *Quat. Sci. Rev.* **28(25-26)** 2675
[75] Anderson RF, Chase Z, Fleisher MQ and Sachs J 2002 *Deep-Sea Res. II* **49** 1909
[76] Crosta X and Shemesh A 2000 *Paleoceanography* **17(1)** doi: 10.1029/2000PA000565
[77] Brzezinski MA et al. 2002 *Geophys. Res. Lett.* **29(12)** doi: 10.1029/2001GL014349
[78] Beucher C, Brzezinski MA and Crosta X 2007 *Global Biogeochem. Cycles* **21** doi: 10.1029/2006GB002746
[79] Cortese G, Gersonde R, Hillenbrand CD and Kuhn 2004 *Earth Planet. Sci. Lett.* **224** 509
[80] Sigman DM, Jaccard SL and Haug G 2004 *Nature* **428** 59
[81] Mahowald NM et al. 2005 *Glob. Biogeochem. Cyc.* **19(4)** doi: 10.1029/2004GB002402
[82] Bareille G, Labracherie M, Bertrand P, Labeyrie L, Lavaux G and Dignan M 1998 *J. Marine Systems* **17** 527
[83] Arrigo KR, van Dijken GL and Bushinsky E 2008 *J. Physics Res.* **C113** doi: 10.1029/2007JC004551
[84] Rose KA, Sikes EL, Guilderson TP, Shane P, Hill TM, Zahn R and Spero H 2010 *Nature* **466** 1093
[85] Rampen SW, Schouten S, Koning E, Brummer G and Sinninghé Damsté J 2008 *Earth Planet. Sci. Lett.* **276** 207
[86] Lamy F, Hebbeln D, Röhl U and Wefer G 2001 *Earth Planet. Sci. Lett.* **185** 369
[87] De la Rocha C, Brzezinski M, DeNiro M and Shemesh A 1998 *Nature* **395** 680
[88] Robinson RS, Brunelle BG, Sigman DM 2004 *Paleoceanography* **19** doi: 10.1029/2003PA000996
[89] Swann GE, Maslin MA, Leng MJ, Sloane HJ and Haug GH 2006 *Paleoceanography* **21** doi: 10.1029/2005PA001147
[90] Shemesh A, Burckle LH and Hays JD 1994 *Science* **266** 1542
[91] Shemesh A, Hodell D, Crosta X, Kanfoush S, Charles C and Guilderson T 2002 *Paleoceanography* **17(4)** doi: 10.1029/2000PA000599

[92] Altabet MA and François R 1994 *Global Biogeochem. Cycles* **8(1)** 103
[93] De la Rocha C, Brzezinski M, DeNiro MJ 1997 *Geochim. Cosmochim. Acta* **61(23)** 5051
[94] Robinson RS, Sigman D, DiFiore PJ, Rohde MM, Mashiotta TA and Lea DW 2005 *Paleoceanography* **20(3)** doi:10.1029/2004PA001114
[95] Hutchins DA and Bruland KW 1998 *Nature* **393** 561
[96] Takeda S 1998 *Nature* **393** 774
[97] Matsumoto K, Sarmiento JL and Brzezinski MA 2002 *Global Biogeochem. Cycles* **16(3)** doi: 10.1029/2001GB001442
[98] Crosta X, Shemesh A, Etourneau J, Yam R, Billy I and Pichon JJ 2005b *Global Biogeochem. Cycles* **19** doi: 3010.1029/2004GB002344
[99] Brunelle B, Sigman D, Cook M, Keigwin L, Haug G., Plessen B, Schettler G and Jaccard S 2007 *Paleoceanography* **22** doi: 10.1029/2005PA001205
[100] François R, Altabet MA, Yu EF, Sigman DM, Bacon MP, Frank M, Bohrmann G, Bareille G and Labeyrie LD 1997 *Nature* **389** 929
[101] Hay MB, Pienitz R and Thompson RE 2003 *Mar. Micropaleontol.* **48** 291
[102] Armand LK, Cornet-Barthaux V, Mosseri J and Quéguiner B 2008 *Deep-Sea Res. II* **55** 563
[103] Dabros M and Mudie P 1986 *Geological Survey of Canada* **86(1A)** 107
[104] Beaufort L, de Garidel-Thoron T, Mix A and Pisias N 2001 *Science* **293** 2440
[105] Schmidt D, Thierstein H and Bollman J 2004 *Palaeogeogr. Palaeoclimatol. Palaeoecol.* **212** 159
[106] Weller A and Girault F Submitted *Mar. Micropaleontol.*
[107] Popp BN, Laws EA, Bidigare RR, Dore JE, Hanson KL and Wakeham SG 1998 *Geochim. Cosmochim. Acta* **62(1)** 69
[108] Lourey MJ, Trull TW and Sigman D 2003 *Global Biogeochem. Cycles* **17(3)** doi: 10.1029/2002GB001973
[109] Varela DE, Pride CJ and Brzezinski MA 2004 *Global Biogeochem. Cycles* **18** doi: 10.1029/2003GB002140
[110] Tortell PD and Morel FMM 2002 *Limnol. Oceanog.* **47(4)** 1012
[111] Karsh KL, Trull TW, Lourey MJ and Sigman DM 2003 *Limnol. Oceanog.* **48(3)** 1058
[112] Sigman DM, Altabet MA, François R, McCorkle DC and Gaillard JF 1999 *Paleoceanography* **14(2)** 118
[113] Cardinal D., Alleman LY, de Jong J, Ziegler K and André L 2003 *J. Anal. Atom. Spectro.* **18** 213
[114] Brzezinski M, Jones J, Beucher C, Demarest M and Berg H 2006 *Anal. Chem.* **78(17)** 6109.
[115] Swann GE, Maslin MA, Leng MJ, Sloane HJ and Haug GH 2006 *Paleoceanography* **21** doi: 10.1029/2005PA001147
[116] Annett AL, Crosta X, Clarke AC and Ganeshram R 2010 *Polar Biology* **33** 13

Dinocysts as tracers of sea-surface conditions and sea-ice cover in polar and subpolar environments

Anne de Vernal[1] and André Rochon[2]

[1] GEOTOP, Université du Québec à Montréal, P.O. Box 8888, succursale "centre ville" Montréal, Qc, H3C 3P8 Canada

[2] GEOTOP and Institut des Sciences de la Mer (ISMER), Université du Québec à Rimouski, 310, Allée des Ursulines, Rimouski, Qc, G5L 3A1 Canada

E-mail: devernal.anne@uqam.ca

Abstract. Dinoflagellates are unicellular protists that produce a cyst (dinocyst) as part of their life cycle. The cyst wall of many species is composed of highly resistant organic matter. Dinocysts are thus routinely recovered in marine sediments and occur in high number along the continental margins of the world oceans notably in high latitude environments. They are widely used as proxy indicators of marine conditions and provide valuable information on the natural variability of climate, which in turn helps understanding and assessing the potential threat posed by the actual global warming. Here we present a brief outline of their biology, ecology and distribution in Arctic and subarctic areas. We also provide a few examples of paleoenvironmental reconstructions and briefly discuss on the significance of these results.

1. Introduction

Dinoflagellates are microscopic unicellular protists belonging to the division of Dinoflagellata [1]. Their populations develop in most types of aquatic environments, from lakes to open ocean, and from equatorial to arctic settings. They constitute an important part of primary productivity in marine waters together with diatoms and coccolithophorids. The living cells of dinoflagellates do not yield remains that fossilize. However, during the course of their life cycle some species produce resistant organic-walled cysts, which protect the diploid cells after fusion of the gametes (figure 1). The cyst permits the survival of cells for periods of variable lengths (from the season to decades). It is composed of refractory organic matter that permits preservation in the sediment and fossilization. The fossil form is usually refereed to as dinocyst.

Dinocysts are routinely recovered in palynological preparations, after treatments involving chemical dissolution of carbonate ($CaCO_3$) and silicate (SiO_2) particles with hydrochloric acid (HCl) and hydrofluoric acid (HF). They are usually well preserved in sediments of continental margins including epicontinental seas, estuaries, continental shelves, slopes and rises. Their concentrations may reach 10^6 cysts per gram [2]. In polar and subpolar seas, dinocyst are usually abundant and their assemblages are characterized by relatively high species diversity.

2. Ecology of dinoflagellates in polar and subpolar seas

Published under licence by IOP Publishing Ltd

The motile forms of dinoflagellates are characterized by the presence of two flagella, one longitudinal and the other located in a depression or cingulum, around the cell. The cell is comprised within a membrane, the amphiesma, which can include a series of cellulosic plates (referred to as "armored" or thecate), while others lack cellulosic plates and are referred to as "naked" or "athecate". About 20% of dinoflagellates produce dinocysts that fossilize as palynomorphs [3]. Most of them are composed of dinosporin, a complex biomacromolecular substance made of phenolic, alcoholic and/or carboxylic hydroxides, and fatty acids accompanied by tocopherols and sterols (mostly cholesterol and dinosterol) [4]. Rare species produce cysts composed of calcium carbonate.

The morphologies of motile and cyst stages of same species are often very different, which makes it difficult to understand the complete life cycle. Also, biologists and marine palynologists independently studied the motile and cyst forms respectively. This gave rise to two distinct nomenclature systems, which further complicates the understanding of their life cycle.

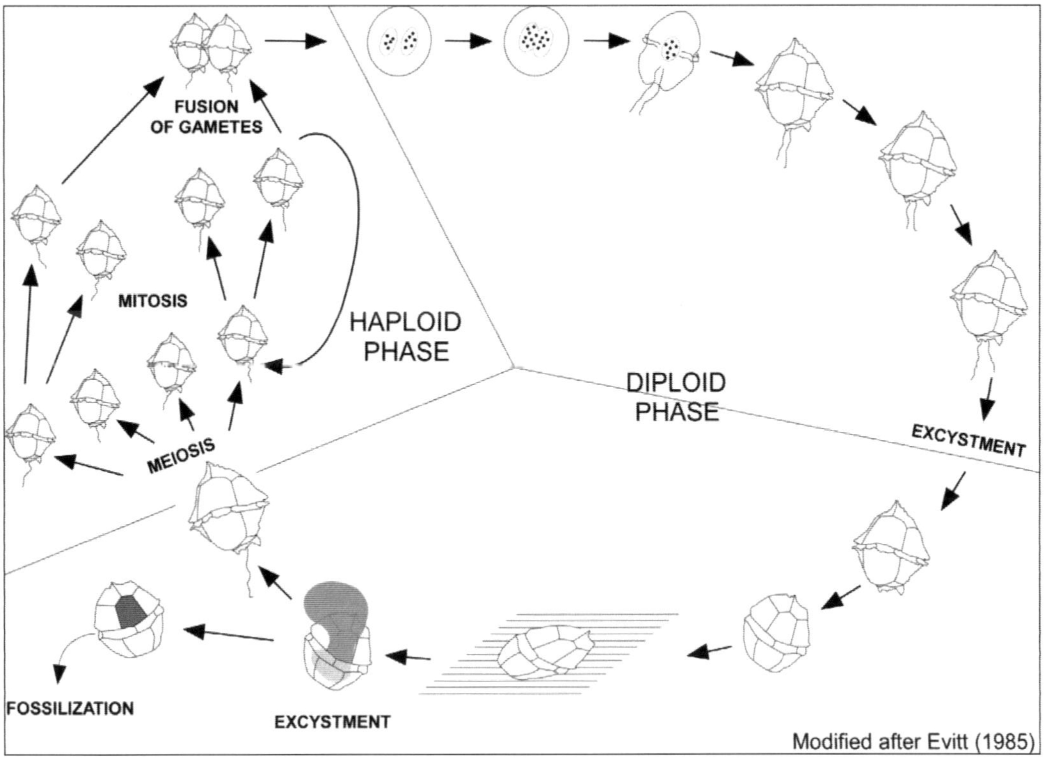

Figure 1. Illustration of the relationship between dinoflagellates and dinocysts.

The distribution of dinoflagellates is not only dependent upon the physical and chemical parameters (currents, temperature, salinity, irradiance, nutrients, sea ice) of the water column, but also on their feeding strategies, which may include mixotrophy and phagotrophy (ingestion of large food particles or preys) as well as the distribution of their prey. Over the last decades, several research programs aimed at studying the changing Arctic environment were conducted in the northern Baffin Bay (International North Water Polynya Study), the Beaufort Sea (Canadian Arctic Shelf Exchange Study) and the Canadian Arctic Archipelago (ArcticNet). The ecology of plankton and ice diatoms and other protists have been extensively studied over long-term periods as part of these Arctic research programs, but very little has been done on the ecology of dinoflagellate bloom events. The main reason is because dinoflagellates are usually much less abundant in the water column than diatoms. Dinoflagellates, which are adapted to growth under low turbulence and low nutrients concentrations, often bloom after diatoms. Moreover, their blooms are usually restricted in time and space, and are therefore easily missed. For example, in highly productive Arctic basins, such as the North Water Polynya in northern Baffin Bay, diatoms usually dominate the phytoplankton and may account for up to 70% of the total phytoplankton carbon biomass, while dinoflagellates and ciliates comprise approximately 24% of the total phytoplankton carbon biomass. However, dinoflagellates may account for the majority of the phytoplankton carbon biomass during certain periods in relation to the depth of the surface mixed layer.

2. Distribution of dinocysts in surface sediments of polar and subpolar seas

Dinocysts are recorded in high concentrations in the Arctic and subarctic seas, but are rare in the central Arctic basin, where the surface waters are characterized by perennial sea ice [5-6]. This is attributed to a relatively high productivity in seasonally ice-free Canadian and Arctic seas contrasting with low dinoflagellate productivity under permanent sea ice. The overall pattern of dinocyst distribution on the sea floor suggests deposition close to the productivity zones as demonstrated by studies from relatively shallow areas (~ 500 m) indicating that the cysts assemblages in sediment are representative of the local/regional dinoflagellate populations in the overlying water column [7-8]. However, between the formation of the cysts in the water column and their settling down to the sea floor, horizontal transport may occur with surface and/or subsurface currents. This implies that cysts retrieved in the surface sediments may have been formed elsewhere and are therefore not necessarily representative of the upper water column conditions immediately above the collection site. Lateral transport is a source of concern especially in offshore areas of low productivity where the dinocyst assemblages are characterized by low concentrations, which makes it difficult interpretations in terms of sea-surface conditions.

Another point to consider is how accurately surface sediment samples represent the modern dinocyst assemblages and the upper water column conditions. Dinocyst abundance in surface sediments may vary according to productivity in the upper part of the water column, but is also dependant on the sedimentation rates. The sediment accumulation rates vary greatly from one region of the Canadian High Arctic to another. Most shelf areas and channels of the Canadian Arctic Archipelago are characterized by low sedimentation rates due to the lack of sediment sources. High sediment accumulation areas are restricted to localized basins. For example, in the Mackenzie Trough the sedimentation rate can reach 0.12 cm per year, while in the nearby Amundsen Gulf, it is of the order of 0.01 mm per year. Therefore, the upper 1 cm of sediment that is routinely collected and analyzed for "modern" dinocyst assemblages may represent between 10 and 10^3 years of sedimentation. This implies that some of the dinocysts recovered from the surface sediments may in fact be indicative of motile dinoflagellate populations that are no longer present in the area. However, this is unlikely for the majority of the dinocyst species found, since repeated surface sediment sampling campaigns over the last 10 years have yielded relatively important concentrations of viable cysts with cell content.

3.1. The most common dinocyst taxa in polar and subpolar seas

In the northern Hemisphere, species diversity of dinocysts is relatively high and comprises about 60 taxa [9-11]. As it is the case for many other organisms, species diversity decreases towards high latitudes and only about 20 taxa are typical of subpolar to polar environments [5-6] (plate 1). Among the taxa that are commonly found in high latitude environments, some have a relatively cosmopolitan distribution (e.g., *Operculodinium centrocarpum*, *Brigantedinium simplex*, *B. cariacoense*). Some are specific to Arctic environments, such as *Polykrikos* var. Arctic and *Polykrikos* var. quadratus, *Islandinium*? *cezare*, *Echinidinium karaense* and *Spiniferites* type *frigidus*). Others are present in low numbers as "accompanying taxa" (e.g., cyst of *Protoperidinium americanum*, cyst of

Pentapharsodinium dalei). One taxon, *Impagidinium pallidum*, has a bipolar distribution and is present in Arctic/subarctic and Antarctic environments.

Morphological variations within a species or between two species have been observed and they sometime make it difficult to assign a specimen to a particular species. *Spiniferites elongatus* and *Spiniferites frigidus* are among these species. An elongated central body and relatively short trifurcate processes characterize them both. However, flanges, called septae, join the base of the processes in *S. frigidus*. Those septae are sometimes present in *S. elongatus*, mostly in the antapical area, but some specimens also have septae joining processes in the cingular area. Therefore we observe a complete morphological gradation between *S. elongatus* and *S. frigidus*, which is the reason why they are grouped as *S. elongatus/frigidus* during routine counts. Another species where morphological variations are often presents is *Operculodinium centrocarpum*, which sometimes displays incomplete (*O. centrocarpum* var. arctic) or short/truncated processes (*O. centrocarpum* short spines or var. *truncatum*) [5].

There are groups of taxa that are sometimes difficult to identify to the species level because specimens are not properly oriented or damaged, or simply because the morphological differences among species are not obvious at first. One of these groups is the *Brigantedinium* spp. group, commonly referred to as "round browns". These can be identified to the species level based on the shape of the archeopyle (the opening through which the dinoflagellate comes out during excystment; see figure 1). If the archeopyle is not visible or properly oriented, the specimen is assigned to *Brigantedinium* sp. Another group in which species are sometimes difficult to identify, even for the well trained palynologist, is the "spiny round brown" cyst group, which includes *Islandinium minutum*, *Islandinium*? *cezare*, *Echinidinium karaense* and *Echinidinium aculeatum*. Only minute morphological details allow differencing one species from another, such as process shape/length/tip or surface ornamentation.

In high latitudes of the southern hemisphere, dinocyst data are available for about 80 surface samples [12]. Among them, some are exclusively reported from the circum Antarctic. This is notably the case of *Selenopemphix antarctica* that is considered "endemic" to the polar areas of the southern hemisphere.

3.2. The relationship between the dinocyst distribution and environmental parameters

In Arctic and subarctic seas, there is a relationship between cyst assemblages and environmental parameters, which include sea ice, productivity, salinity and temperature. In polynyas, which are characterized by high primary productivity, cysts of heterotrophic dinoflagellate largely dominate the assemblages [13]. The almost exclusive occurrence of heterotrophic taxa in polynyas is also a characteristic of many upwelling areas. It is interpreted as the consequence of productivity dominated by diatoms, which have the ability to multiply much more rapidly than dinoflagellates [11]. In contrast to low latitude upwelling, however, the dominance of Protoperidiniales belonging to *Brigantedinium* spp. and *Islandinium minutum* and the occurrence of *Echinidinium karaense* are characteristic features.

Multivariate analyses performed on the Northern Hemisphere database that includes 1171 sites [11] (see data base at http://www.GEOTOP.ca) demonstrate that the dinocyst distribution is controlled by several parameters, notably surface salinity, winter and summer temperature, productivity and sea ice (figure 2a). When the polar areas north of 66°N are isolated (n = 401 sites), the seasonal duration of sea-ice cover stands as the most determinant parameter (figure 2b).

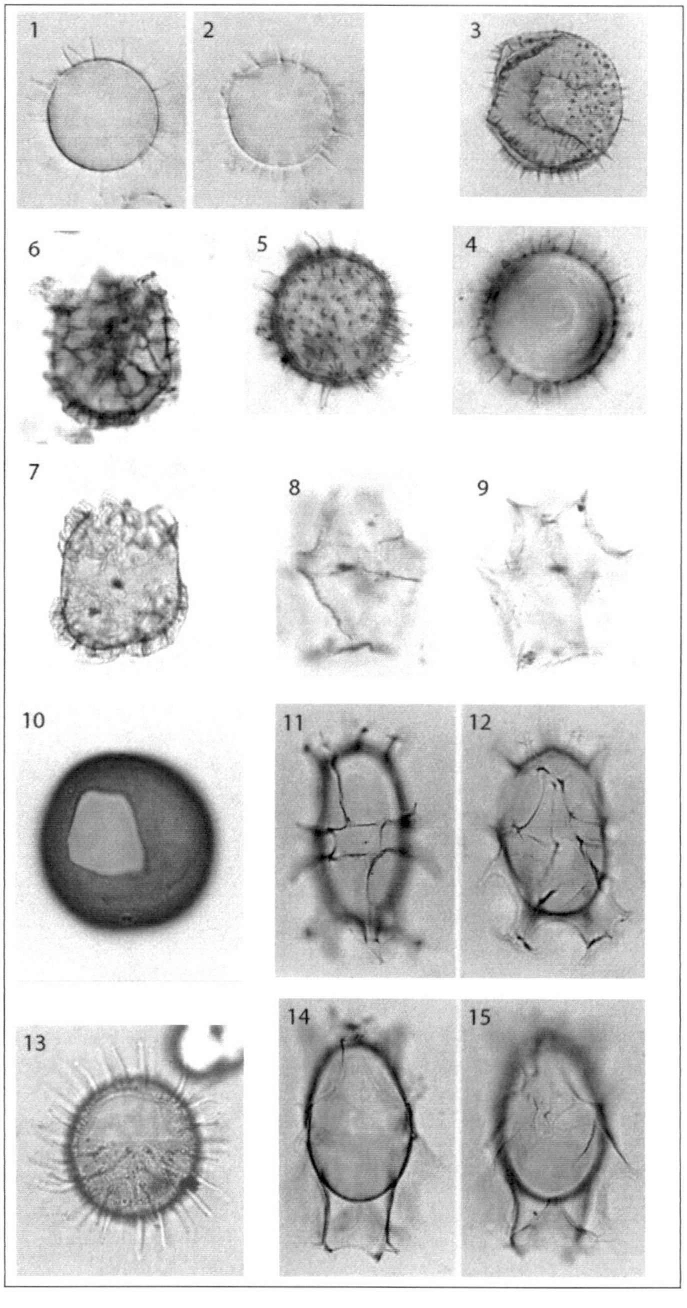

Plate 1. Photographs of the most common arctic and subarctic dinocyst taxa : (1-2) *Pentapharsodinium dalei*, (3) *Islandinium minutum*, (4) *Echinidinium karaense*, (5) *Islandinium*? *cezare*, (6-7) *Polykrikos* sp. var. arctic and *quadratus*, respectively, (8-9) *Impagidinium pallidum*, in dorsal and ventral views, respectively), (10) *Brigantedinium simplex*, (11-12) *Spiniferites elongatus* in dorsal and ventral views respectively, (13) *Operculodinium centrocarpum*, (14-15) *Spiniferites frigidus*. For taxonomy and comments on the morphology and distribution of these taxa, see references [5], [9] and [14].

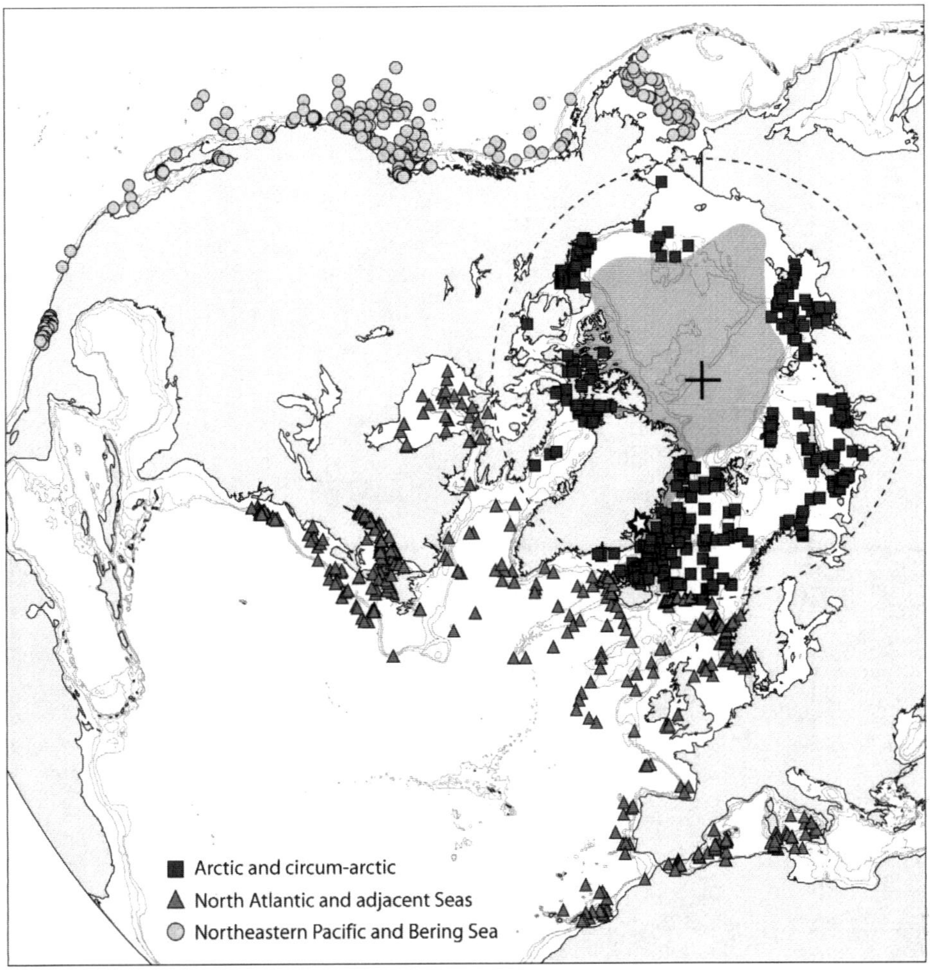

Figure 2a. Map showing the location of surface sediment samples in the reference dinocyst database (1171 sites and 64 taxa). Reprinted from [11] with permission from Elsevier. The blue squares correspond to Arctic sites, north of 66°N. The red triangles and green points correspond to Atlantic and Pacific sites, respectively. The minimum sea-ice cover extent is shown by the gray zone. No sites under perennial sea-ice are included inasmuch as the assemblages are barren in such conditions due to extremely low productivity. Isobaths correspond to 200 m, 1000 m and 2000 m.

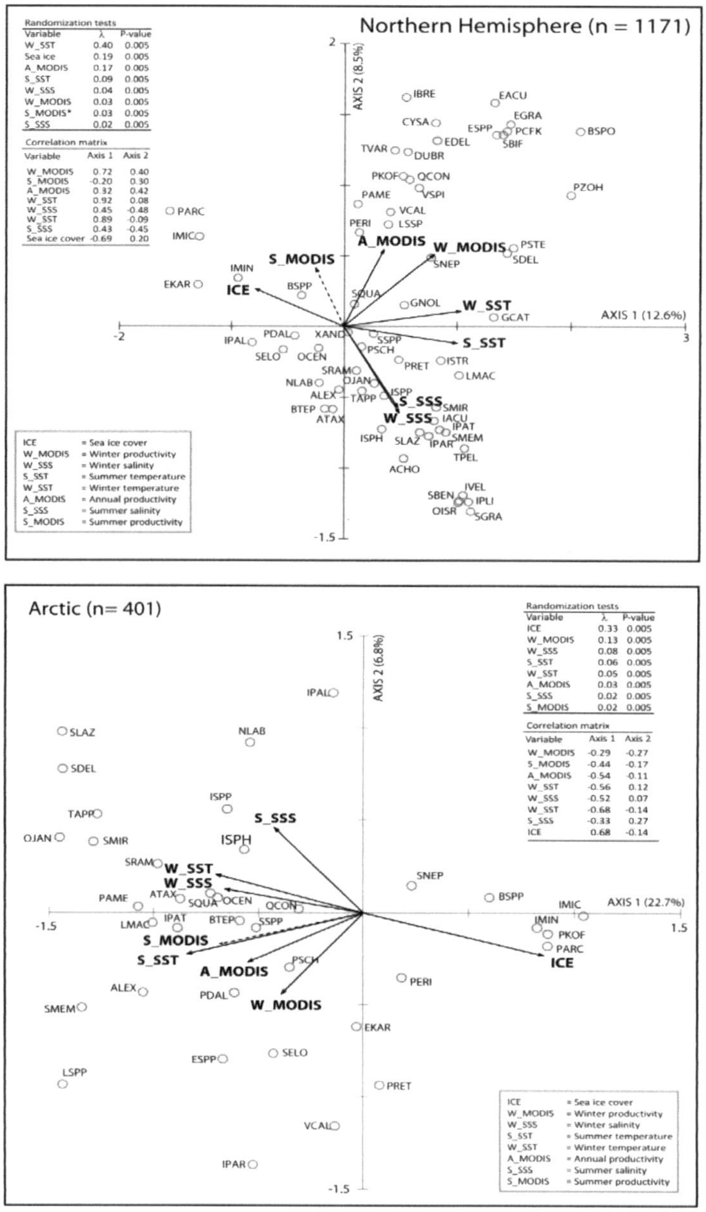

Figure 2b. Results of canonical correspondence analyses performed on the dinocyst assemblages (64 taxa) and surface ocean parameters (sea-ice, temperature, salinity, productivity) using the Canoco software [15]. Sea-ice stands as a determinant parameter on the assemblage distribution at the scale of the northern hemisphere (upper panel) and as the most determinant parameter (lower panel) north of 66°N (blue squares in figure 3a), which includes 401 sites and 37 dinocyst taxa (figure modified from [11]; data and codes of dinocyst taxa are provided in the database archived on the GEOTOP website).

4. Using dinocysts for quantitative reconstructions of sea-ice cover, sea-surface temperature and salinity

Transfer functions were developed for the reconstruction of past environmental conditions based on microfossil assemblages. The relationships between environmental parameters and dinocyst assemblages are complex (non linear) and differ from one basin to another (from the Atlantic to the Arctic). Thus, simple transfer function based on calibrations cannot be applied unequivocally and the modern analogue technique (MAT) is the preferred method used so far [16-17]. Validation tests were made to evaluate the performance of the approach and to calculate the error of prediction (see table 1). The reliability of the approaches is given by the coefficient of correlation (r^2) between observed and estimated values, whereas the reproducibility is provided by the root mean square error (RMSE) that corresponds to the standard deviation of the difference between observed and estimated value. In the case of the Northern Hemisphere database, the error of the prediction is close to the range of inter-annual variability for most parameters, including sea-ice cover.

Most studies using paleoceanographical reconstructions based on dinocysts followed the same procedures [18]. This procedure includes the log transformation of taxa percentages prior to search for analogues in the reference database. Such a transformation is indeed necessary to emphasize the weight of accompanying taxa, which often have narrow ecological affinity whereas the most abundant taxa are often cosmopolitan.

Table 1. Accuracy of past sea-surface condition estimates based on MAT applied to dinocyst assemblages. The results are from the Northern Hemisphere database (n = 1171 sites and 64 taxa; data available in paleoceanographic data pages under the GEOTOP webpage (www.GEOTOP.ca). "r^2" refers to the coefficient of correlation between observed and estimated values and the RMSE is the error in the reconstruction, which corresponds to the standard deviation (1sigma) of the difference between observed and estimated values.

Parameter	**Standard deviation (1 sigma) Instrumental data**	r^2	**RMSE**
Winter temperature	±1.08 °C	0.97	1.19 °C
Summer temperature	± 1.55 °C	0.95	1.71 °C
Summer salinity (>30)	± 0.63	0.9	0.62
Sea-ice cover	± 1.19 months a^{-1}	0.89	1.24 months/yr
Winter productivity	na	0.92	11.5 gCm^{-2} (~ 20%)
Summer productivity	na	0.79	23 gCm^{-2} (~ 25%)
Annual productivity	na	0.84	55 gCm^{-2} (~ 18%)

5. Examples of paleoceanographical reconstructions from dinocysts in polar seas

5.1. Example from the Canadian Arctic

An east-west transect of 6 sediments cores were collected in the Canadian High Arctic and analyzed for their content in dinocysts. Quantitative reconstructions of sea surface parameters were performed on each core in order to characterize the evolution of the marine environment throughout the Holocene time-period (the last 11 000 years). The evolution of sea-surface temperature is illustrated on figure 3 and presented as the departure from the modern conditions.

Maximum temperature, up to 3°C higher than at present, occurred around 6000 BP in Jones and Lancaster Sounds, and around 5000 BP in Barrow Strait. This is what is usually referred to as the Holocene thermal optimum for this region. These 3 locations are also characterized by a slight cooling trend that continues until recently. The Dease Strait, in the central Canadian Arctic Archipelago, was characterized by a series of low amplitude warm/cold oscillations for the last ~8000 years without marked temperature maximum. The western Arctic sites of the Mackenzie Slope and Trough are characterized by a slight warming trend that started around 9000 BP. These reconstructions illustrate the opposite temperature trends that occurred in the Arctic since about 11 000 years: the Eastern Canadian Arctic is recording cooling after a temperature maximum around 6000-5000 BP, while the western Arctic records warming. We tentatively associate such climate trend with large-scale atmospheric patterns involving positive mode of Arctic Oscillation during the early Holocene [19]. More local conditions such as coastal currents, the halocline structure or the proximity of an ice-cap (e.g., the Inuitian Ice Sheet in the eastern Canadian Arctic) probably played an important role on the sea-surface temperature and sea-ice cover, being thus responsible for strong regionalism.

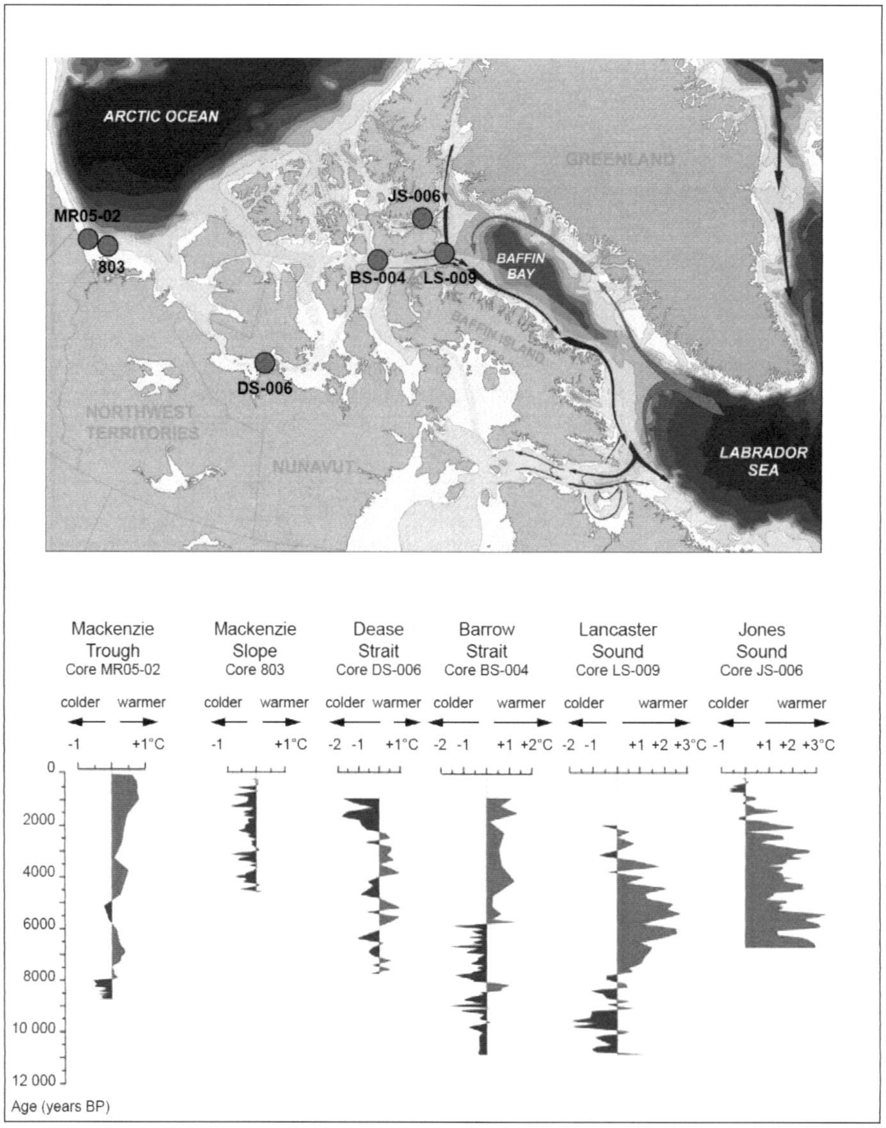

Figure 3. Sea-surface temperature reconstructions from cores collected along an East-West transect in the Canadian high Arctic (location shown in the upper panel). Data are expressed as departure from modern day conditions (lower panel) [20]. The Holocene thermal optimum is well represented in the eastern cores (Jones and Lancaster Sound and Barrow Strait).

5.2. The early Holocene in circum-Arctic regions

Quantitative Holocene sea-surface reconstructions based on dinocysts are available in a number of cores from the northern North Atlantic and adjacent subpolar seas (see figures 3 and 4). The results show a strong regionalism in sea-ice distribution, even more pronounced than at present. At some sites, data reveal a less extensive sea-ice cover around 8000 years BP than during the late Holocene. This is the case of the eastern Fram Strait, northern Baffin Bay, Hudson Bay and St. Lawrence Estuary [19-20]. However sites located in the western Arctic [21], along the Labrador [22] and eastern Greenland [23] margins recorded a denser sea-ice cover.

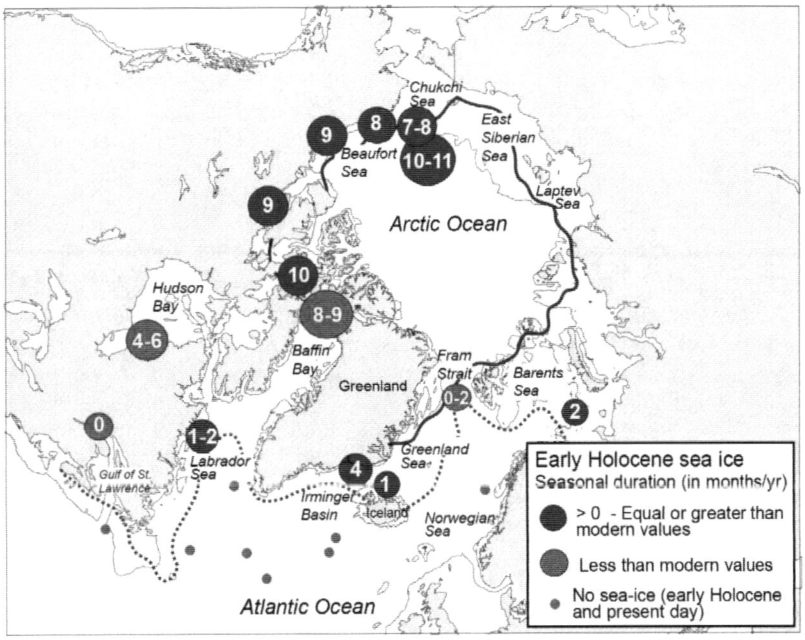

Figure 4. Seasonal duration (months/yr) of sea ice at 8000-7000 years before present based on paleoclimate reconstructions from the analyses of the microfossil content of sediment cores [19-20]. The thick and dashed lines correspond to the median of September sea-ice extent from 1979 to 2000 and the maximum sea-ice extent in winter, respectively. The results demonstrate dense sea-ice cover in the Canadian Arctic even during the warmer part of the Holocene. They also illustrate different seasonal distribution pattern than at present.

5.3. Pleistocene dinocyst records from the Arctic and circum-Arctic

Pleistocene records of dinocysts assemblages are available for a number of sites in subarctic areas, such as the Gulf of Alaska [24], the Labrador Sea [25] and the Russian Arctic [26]. They illustrate large amplitude variation in assemblages with higher concentrations and species diversity during interglacial than during glacial stages. For what concerns the central Arctic Ocean, however, no record is available. Analyses of cores collected during HOTRAX and ACEX expeditions in the central Arctic did not yield high enough number of dinocyst specimens to describe the assemblages (unpublished observation). Taking into account low sedimentation rates, the rarity of dinocysts led to conclude that extremely low dinocyst fluxes characterized the central Arctic basin during the Pleistocene, likely due to low productivity because of a perennial sea-ice cover. Such a conclusion, however, should be taken with caution as it is based on negative evidence.

6. Conclusion

Dinocysts are extremely useful in high latitude paleoceanography and present several advantages over other microfossil proxy indicators. The chemical composition of their membrane allows them to be preserved in sediments where siliceous (e.g., diatoms) and calcareous (e.g., foraminifers and coccoliths) organisms may suffer dissolution. The relatively high species diversity in polar and subpolar environments make them useful tracers of sea-surface conditions in cold settings, as opposed to other groups of organisms, such as planktonic foraminifers, which form monospecific assemblages when sea-surface temperatures are below 5°C. Their use as indicators of sea-surface conditions using transfer functions has provided paleoceanographical information such as sea-ice cover that is not accessible with other indicators. This not only helps understanding climate variability at millennial and secular timescales, but the reconstructed sea-surface parameters are also used as boundary conditions in climate modeling experiments, which contribute to predict future climate changes in the context of the actual global warming.

References

[1] Fensome R A, Taylor F J R, Norris G, Sarjeant W A S, Wharton D I and Williams G L 1993 *A classification of living and fossil dinoflagellates* (*Micropaleontology special publication* n°7) (New York: American Museum of Natural History) p 351
[2] de Vernal A, Rochon A and Radi T 2007 Dinoflagellates *Encyclopedia of Quaternary Sciences* ed S Elias (Amsterdam: Elsevier) pp 1653–69
[3] Dale B 1976 Cyst formation, sedimentation and preservation: factors affecting dinoflagellate assemblages in recent sediments from Throndheimsfjord, Norway *Rev. Palaeobot. Palynol.* **22** 39–60
[4] Kokinos J P, Eglinton T I, Goni M A, Boon J J, Martoglio P A and Anderson D M 1998 Characterization of a highly resistant biomacromolecular material in the cell wall of a marine dinoflagellate resting cyst *Org. Geochem.* **28** 265–88
[5] de Vernal A, Henry M, Matthiessen J, Mudie P J, Rochon A, Boessenkool K, Eynaud F, Grøsfjeld K, Guiot J, Hamel D , Harland R, Head MJ, Kunz-Pirrung M, Levac E, Loucheur V, Peyron O, Pospelova V, Radi T, Turon J-L and Voronina E 2001 Dinoflagellate cyst assemblages as tracers of sea-surface conditions in the northern North Atlantic, Arctic and sub-Arctic seas: the new "n = 677" database and application for quantitative paleoceanographical reconstruction *J. Quat. Sci.* **16** 681–698
[6] Matthiessen J, de Vernal A, Head M, Okolodkov Y, Zonneveld K and Harland R 2005 Modern organic-walled dinoflagellate cysts in Arctic marine environments and their (paleo-) environmental significance *Paläontologische Zeitschrift* **79/1** 3–51
[7] Anderson D M, Stock C A, Keafer B A, Nelson A B, McGillicuddy D J, Keller M, Thompson B, Matrai P A and Martin J 2005 *Alexandrium fundyense* cyst dynamics in the Gulf of Maine *Deep-Sea Res. II* **52** 2522–42
[8] Reid P C 1975 A regional subdivision of dinoflagellate cysts around the British Isles *New Phytol.* **75** 589–603
[9] Rochon A, de Vernal A, Turon J-L, Matthiessen J and Head M J 1999 *Distribution of recent dinoflagellate cysts in surface sediments from the North Atlantic Ocean and adjacent seas in relation to sea-surface parameters* (College Station TX: American Association of Stratigraphic Palynologists Foundation) *Contribution Series Number* **35** 152 p
[10] Marret F and Zonneveld K A F 2003 Atlas of modern organic-walled dinoflagellate cyst distribution *Rev. Palaeobot. Palynol.* **125** 1–200
[11] Radi T and de Vernal A 2008 Dinocysts as proxy of primary productivity in mid-high latitudes

of the Northern Hemisphere *Mar. Micropaleontol.* **68** 84–114

[12] Marret F, de Vernal A, Benderra F and Harland R 2001 Reconstruction of late Quaternary sea-surface conditions at DSDP Site 594 in the southwest Pacific Ocean based on dinoflagellate cyst assemblages *J. Quat. Sci.* **16** 739–49

[13] Hamel D, de Vernal A, Gosselin M and Hillaire-Marcel C 2002 Organic-walled microfossils and geochemical tracers: sedimentary indicators of productivity changes in the North Water and northern Baffin Bay (High Arctic) during the last centuries *Deep-Sea Res. II* **49** 5277–95

[14] Head M J, Harland R and Matthiessen J 2001 Cold marine indicators of the late Quaternary: the new dinoflagellate cyst genus *Islandinium* and related morphotypes *J. Quat. Sci.* **16** 621–36

[15] Ter Brack C J F and Smilauer P 1998 *Canoco reference manual and user's guide to Canoco for Windows, software for canonical community ordination* (version 4) (Wageningen: Centre for Biometry) 351 p

[16] Guiot J and de Vernal A 2007 Transfer functions: methods for quantitative paleoceanography based on microfossils *Proxies in Late Cenozoic Paleoceanography (Developments in Marine Geology* vol 1*)* ed C Hillaire-Marcel and A de Vernal (Amsterdam: Elsevier) chapter 13 pp 523–63

[17] Bonnet S, de Vernal A, Hillaire-Marcel C, Radi T and Husum K 2010 Variability of sea-surface temperature and sea-ice cover in the Fram Strait over the last two millennia *Mar. Micropaleontol.* **74** 59–74

[18] de Vernal A, Eynaud F, Henry M, Hillaire-Marcel C, Londeix L, Mangin S, Matthiessen J, Marret F, Radi T, Rochon A, Solignac S and Turon J-L 2005 Reconstruction of sea-surface conditions at middle to high latitudes of the Northern Hemisphere during the Last Glacial Maximum (LGM) based on dinoflagellate cyst assemblages *Quat. Sci. Rev.* **24** 897–924

[19] de Vernal A, Hillaire-Marcel C, Solignac S, Radi T and Rochon A 2008 Reconstructing sea-ice conditions in the Arctic and subarctic prior to human observations *Arctic Sea ice Decline: Observations, Projections, Mechanisms, and Implications* ed E T DeWeaver, C M Bitz and L B Tremblay (Washington DC: AGU Monograph Series) **180** 27–45

[20] Ledu D, Rochon A, de Vernal A, Barletta F and St-Onge G 2010 Holocene sea ice history and climate variability along the main axis of the Northwest Passage, Canadian Arctic *Paleoceanography* **25** PA2213 doi:10.1029/2009PA001817

[21] de Vernal A and Hillaire-Marcel C 2006 Provincialism in trends and high frequency changes in the northwest North Atlantic during the Holocene *Global Planet. Change* **54** 263–90

[22] McKay J L, de Vernal A, Hillaire-Marcel C, Not C, Polyak L and Darby D 2008 Holocene fluctuations in Arctic sea-ice cover: Dinocyst-based reconstructions for the eastern Chukchi Sea *Can. J. Earth Sci.* **45** 1377–97

[23] Solignac S, Giraudeau J and de Vernal A 2006 Holocene sea-surface conditions in the western Nordic Seas: spatial and temporal heterogeneities *Paleoceanography* **21** PA2004 doi: 10.1029/2005PA001175

[24] Marret F, de Vernal A, Pedersen T F and McDonald D 2001 Middle Pleistocene to Holocene palynostratigraphy of ODP 887 in the Gulf of Alaska, northeastern North Pacific *Can. J. Earth Sci.* **38** 373–86

[25] de Vernal A and Mudie P J 1992 Pliocene and Quaternary dinoflagellate cyst stratigraphy in Labrador Sea: paleoecological implications *Neogene and Quaternary dinoflagellate cysts and acritarchs* ed M J Head and J H Wrenn (College Station TX: American Association of Stratigraphic Palynologists Foundation) 329–46

[26] Matthiessen J, Knies J, Nowaczyk N and Stein R 2001 Late Quaternary dinoflagellate cyst stratigraphy at the Eurasian continental margin, Arctic Ocean: indications for Atlantic water inflow in the past 150 000 years *Global Planet. Change* **31** 65–86

IOP Conf. Series: Earth and Environmental Science **14** (2011) 012008 doi:10.1088/1755-1315/14/1/012008

Transfer functions

Joel Guiot

CEREGE/ECCOREV, CNRS – Aix-Marseille Université, Europôle de l'Arbois BP80, 13545 Aix-en-Provence Cedex 04, France

E-mail: guiot@cerege.fr

Abstract. This chapter summarizes the different approaches used to reconstruct past environmental conditions from micropaleontological proxy data. The first group of methods involves calibration, i.e. a relationship is calculated between a given environmental variable and the abundances of the various species (e.g. regression, partial least squares). The second group involves the calculation of a similarity index between the assemblages (modern analogue technique on raw assemblages or based on surface responses). The validation of the transfer functions must be done using several well known statistics calculated on independent modern data (coefficient of determination, root-mean-square error ...). In conclusion, limitations and potentials of the different approaches are discussed.

1. Introduction

Transfer functions refer to a set of statistical methods used in paleoceanography and in palaeoclimatology in general to express microfossil assemblages in terms of climatic or oceanographic parameters. In paleoceanography, the pioneers are Imbrie and Kipp [1], who reconstructed their reconstruction of sea-surface temperatures based on multivariate analyses of planktonic foraminifer assemblages. Transfer function models and methods based on calibrations involve a few assumptions: (1) environmental variable is the ultimate cause of changes in the paleobiological data; (2) the ecological properties of the species considered has not changed between the period analyzed and the present time, and the relationship between the species and the environment is thus uniform through time; (3) the modern observations contain all the necessary information to interpret the fossil data. The second assumption originates from the uniformatarian principle. It is implicit in most paleoecological studies, but constitutes an unprovable premise, especially when going back in time in the geological archives.

2. Methods based on calibration

The transfer function is a predictive method and is the inverse of the response function. Whereas response functions aim at describing the relationships between species and abiotic factors, transfer function models aim at making predictions that are as close as possible to reality. So climatic variables are predictands (variables to be predicted) and species are predictors (sometimes called independent variables). The error of prediction is, however, difficult to evaluate and is never known with certainty. It is generally larger than the calibration error, which is obtained by comparing the data used for calibration (also called "training data set") and the estimates. The prediction error can be evaluated by a validation procedure in which part of the modern data available is not used for the calibration, and is reserved for prediction tests.

Most of the techniques used are based on ordination of the assemblages. Ordination is a general term for multivariate techniques that sort multi-dimensional data points along one or several axes in order to visualize their structure. In paleoceanography, we hope that this order reflects ordered along the climatic gradients that we want to reconstruct. There are different ordination techniques, all of which differ slightly with respect to the mathematical approach used to calculate species and sample similarity/dissimilarity (Table 1) [2].

Table 1. Classification of the different methods. Generalized models, Non-parametric methods and presence/absence methods are not presented in the text.

Type of approach	Family	Sub-family	Method	Order
Calibration	Ordination techniques	Indirect gradient analyses	Regression on principal components (e.g., the Q-mode Imbrie and Kipp method - **I&K**)	Linear or quadratic
			Regression on correspondance factors (Roux, 1979)	Unimodal
		Direct gradient analyses	Partial least Square (PLS; ter Braak et al., 1993)	Linear
			Canonical correlation (e.g., Guiot et al., 1983)	Linear
			Canonical correspondance analyses (e.g., Gasse and Tekaia, 1983)	Unimodal
			Weighted averaging - Partial least-squares (**WA-PLS**; e.g., Korkhola and Weckström, 2000)	Unimodal
	Generalized models (**GLM**)		Logistic regression (e.g., ter Braak and Looman, 1986	Unimodal
	Non-parametric methods		Generalized additive model (**GAM**; e.g., Gersonde et al., 2005)	Non-linear
			Artificial neural network (**ANN**; e.g., Malmgren and Nordlund, 1996)	Non-linear
Similarity	Presence/ absence		Indicator species method (ISM; Iversen, 1944)	Non-linear
			Mutual climatic range method (MCR; Atkinson et al., 1987)	Non-linear
	Abundances	Unconstrained	Modern analogue technique (**MAT**;Hutson, 1980)	Non-linear
		Constrained	MAT constrainted with external variable (e.g., Guiot et al., 1993)	Non-linear
			MAT constrained geographically (SIMMAX; Pflaumann et al., 1996)	Non-linear
		Response surface	Response surface method (Bartlein et al., 1986)	Non-linear
			Revised analogue method (**RAM**; Waelbroeck et al., 1998)	Non-linear

2.1. Principal component analysis

The Principal Components Analysis (PCA) is one of the earliest ordination techniques and is the basis of the Imbrie and Kipp method (I&K). It uses a rigid rotation to derive orthogonal axes, which maximizes the variance in the species data set. However, it does not guarantee that the variance of the climatic variables is maximized. It is then accompanied by a multiple regression to express these species-defined axes in terms of climate variables.

Imbrie and Kipp [1] have developed a procedure in three steps. The first step consists of describing the biotic assemblages in core-tops in term of hydrographic variables using a PCA, which is an objective way of clustering different species into assemblages of species. The PCA used is a Q-mode on non-centered percentages. A small number of components usually represents a large proportion of the variance. The components retained are related to typical assemblages, such as arctic, subarctic, subtropical and tropical. The second step leads to the derivation of a relationship between each environmental variable and these components using multiple (linear or quadratic) regressions (possibly stepwise regression). The third step projects the down-core data on the principal components and thus allows for the calculation of factor scores for each sample and finally, the scores are entered into the calibration equations to calculate sea-surface temperatures.

CLIMAP [3] have adopted this method in their worldwide reconstruction of the last glacial maximum (LGM) sea-surface temperature (SST). More recently, Gersonde et al. [4] have also used this technique, among two others, to estimate the climate at the LGM in the Circum-Antarctic Ocean based on siliceous microfossils.

2.2. Partial least square (PLS) analysis and canonical correlation analysis

While the PCA does not try to fit at best the predictand, partial least square analysis (PLS) maximizes the covariance between the predictand and a linear combination of the species subjected to the condition that the coefficients of this linear combination are normalized to 1. The analysis aims at finding a linear combination to maximize the covariance with the environmental parameter. As in PCA, the following components must be orthogonal to the first one and together. In this method, the interpretation of the components is directly linked to the predictand to be reconstructed, whereas in the PCA, the components must be calibrated *a posteriori* on the predictor (a predictor is a variable used to predict an dependent variable, they are here the taxa percentages).

2.3. Correspondence analysis

With species assemblages, correspondence analysis (CA) is often more suitable than PCA because it models unimodal relationships between species and climate parameters [5]. It is also an indirect gradient method and, as in the PCA method, the predictand is estimated by regression on the factors. Roux [6] applied this method to the dataset used by Imbrie and Kipp [1] and showed it to perform better.

2.4. Weighted averaging partial least square regression (WA-PLS)

The WA-PLS method extends the CA method in the same way that PLS extends PCA [7]. A particular case of this analysis is the weighted averaging method (WA) where only one component is retained. This method has been widely used to reconstructing chemical parameters from lake diatoms, but has also been applied in paleoceanography [8].

3. Methods based on similarity

An alternative to the transfer function model is the analogue-based approach, introduced for the first time in marine sciences by Hutson [9]. Here, there is no strict calibration. Instead, the assumption is made that a given assemblage of taxa in the fossil record is most likely to have occurred under a combination of environmental conditions that characterize similar modern assemblages of taxa. This approach therefore relies on the comparison between assemblages rather than the direct relationship between taxa and climate or hydrographic parameters. The mutual climatic range methods (and their predecessor, the species indicator method) use maps of the modern distribution of species and consider only the occurrence in terms of presence/absence. We focus here on methods based on abundances.

3.1. Modern-analogues method

The modern analogue technique (MAT) requires a large number of modern or surface samples representing the same type(s) of sedimentary environment as the fossil samples from which microfossil assemblages have been recorded. An appropriate distance measure is required in order to evaluate the degree of analogy between fossil and modern samples [10]. It is possible to use a variety of distance measures with percentage data, especially with data transformations reducing the influence of the ubiquitous taxa that are often the more abundant in the assemblages. The most widely used distance measure is known as the "chord distance", which is particularly effective with percentage data.

The smaller the chord distance is, the greater the degree of analogy between the two samples. Paleoenvironmental reconstructions are based on a small number of modern samples that are the most similar to the fossil sample. The dispersion of the analogues around the reconstructed value provides a partial estimate of the uncertainty, which is related to the tolerance of the proxy in a relatively large range of climatic or oceanographic conditions. However, this estimate does not include any uncertainty related to the fitting error between the climatic or oceanographic conditions. Another issue in estimating the uncertainty depends upon the number of similar assemblages available for a given sample. A way to remedy this problem is to do an even re-sampling of the modern assemblages similar to that of the response surface approach (cf. Section 3.2). The determination of the number of analogues to be taken is also critical. Waelbroeck et al. [11] proposed selecting the set of analogues before a sharp increase of the distance; Gavin et al. [12] proposed another statistical tool named the "receiver operating characteristic" curve which helps to decide the threshold between analogue and no-analogue spectra. De Vernal et al. [13] have shown its ability to reconstruct past SST, salinity and ice cover in subpolar seas from dinocyst assemblages.

3.2. Response surfaces and revised analogue method (RAM)

The first step of the response surface analogue technique (RSAT) is a re-mapping of the taxa into an environmental space using the modern data set. The most common used method of this type currently applied is the quantitative environmental response surfaces [14]. In this method, surfaces are fitted to the abundance values of each taxon used for reconstructions. Having fitted response surfaces for the suite of taxa used in reconstructions, the values for the taxa at each grid point can be combined to provide a series of assemblages that may be considered as new potential (and virtual) analogues. Thereafter, the procedure is very similar to that of MAT, with the 5 to 10 closest matching assemblages selected for each fossil assemblage and used to provide a reconstruction and a confidence interval. The degree of analogy is once again conventionally measured using the chord distance.

In the RSAT approach, the variation of taxa that have no or poor relationship to the climate parameter of interest is smoothed and thus removed by the procedure used to fit the surface. In such cases, the response surface is flat and the taxon contributes little to the choice of analogues. As the new potential assemblages are sampled on a grid, each type of climate is evenly represented, which improves the value of the confidence intervals. These two features represent the greatest advantages of this method. In paleoceanography, the method proposed by Waelbroeck et al. [11], the revised analogue method (RAM), uses gridded assemblages in addition to raw assemblages. The use of RAM allows for improvements in the reconstruction of SSTs at cold and warm ends of the temperature range.

4. Discussion

There is no really conclusion to draw from all these approaches. Guiot and de Vernal [2] have compared the different methods with an example and showed weaknesses and advantages of each of them. We prefer summarize here a few critical points, which concern all of them.

4.1. Validation

To assess the uncertainties associated with the various techniques, many studies use the root-mean-square error (RMSE). This is the square-root of the mean of the squared differences between the observed and the predicted values divided by the number of observations. This statistic is applied to the data used for calibration and it has no power to evaluate the ability of the method when applied to new data, e.g. when reconstructing down-cores. To really assess the extrapolative ability of the method, it is necessary to reserve an independent dataset (not used in the calibration or as the reference dataset of the MAT). When the RMSE is applied to this verification set, it is called the root-mean-square error of prediction (RMSEP). The squared RMSE divided by the variance of the predictand (on the calibration data) is equal to $1\text{-}R^2$ (R^2 is the determination coefficient); the squared RMSEP divided by the variance of the predictand on the verification data is equal to $1\text{-}R_P^2$.

Statistics calculated on the calibration data are simply a measure of the ability of the method to fit the training data set, while those computed on the verification data really assess the predictive capability of the model. As models become more complex (e.g. large number of predictors, strong non-linearities), the R^2 increases even if there is no better correlation between the predictand and their estimates. At the limit, when the number of predictors is equal to the number of observations, R^2 is equal to 1 even with random data. A spurious model with a high R^2 has necessarily a R_P^2 close to zero. However, the verification dataset must be chosen with care. If the verification data set contains samples that are located geographically too close to the samples of the training data set, full independency of the data might not be achieved, which results in an overestimation of the quality of the reconstruction. We recommend to use a verification dataset sufficiently distant (in a geographical sense) from the calibration dataset, but not too climatically different. If we reserve a dataset at the cold margin of the modern dataset, we are able to test if the method applies for glacial conditions, but by doing that we loss good analogues for these conditions. This was done in [2] with the aim of performing such extreme tests, but the final reconstruction must be based on as large of modern data as possible, covering a broad range of environmental conditions.

Examples of RMSEP are given in [2]. When tested on modern planktonic foraminifer data in the North Atlantic, the best methods produce temperature uncertainties of about 1.3°C, and those performing less well produce uncertainties of 2.2°C. In this case, any model that has a mean error of 1.5°C should be considered extremely satisfactory since the actual standard deviation of the "modern" mean sea-surface temperature is >1.5°C in the largest part of the World Ocean.

4.2. How many taxa can be used?

Racca and Prairie [15] have discussed the selection of relevant taxa for diatom-based transfer functions in paleolimnology, which is a particularly delicate matter due to the extremely large diversity of diatom species and the large number of potentially determining environmental parameters in lakes. Generally speaking, some methods are more sensitive to large numbers of predictors. This is related to the parsimony principle, which stipulates that among several equivalent models, it is always better to choose the one having the lowest number of parameters. The first rule to follow is to remove the species that have been shown to have little or no relation to the variables to be reconstructed. The second rule is to preferably use orthogonal factors (principal components or correspondence factors) to restrict the number of parameters (this is preferable to the stepwise selection of predictors).

4.3. Multi-method approach

A multi-method approach has been proposed to obtain more robust reconstructions and to identify the assemblages where transfer functions perform less well. Guiot and de Vernal [2] have shown that it is not necessary to try all of the different methods, as some clearly under-perform and others yield quite similar results. The best candidates for a multi- method approach are MAT, RAM and WA-PLS. This leaves the problem of how to choose between the various reconstructions. Should we select one, which is assumed to be the "best", or average all estimates? Averaging several reconstructions, however, may be biased by the inclusion of outliers, thus leading to an average different from the initial reconstructions. We can advise researchers to make at least three reconstructions, eventually remove possible outliers, and, finally, to use the median reconstruction, which is less sensitive than the mean to outliers. Brewer et al. [16] used the dispersion of the results given by eight methods to estimate the uncertainties associated to the reconstruction method.

4.4 Correlated environmental variables

Another classical problem of transfer functions comes from the possible correlation between the environmental variables to be reconstructed. It is necessary to collect surface samples representative of diverse environmental conditions in order to limit this problem, otherwise it is difficult to make an inference about a given parameter if conditions of the past were different from the present, and the nature of these correlations had changed. Another strategy to evaluate is to select an appropriate subset of species for each environmental variable analyzed. Such a "stepwise" approach based on different predictors may help in producing calibration equations for each predictand that are more independent. Finally we recommend reconstructing the climate variables which are directly related to the proxies and to avoid reconstructing variable indirectly correlated to the proxies as this chain of correlation may have changed in the past. Understanding the functioning of the proxy is the key of a good reconstruction.

References

[1] Imbrie J and Kipp N G 1971 A new micropaleontological method for quantitative paleoclimatology: application to a late Pleistocene Caribbean core *The Late Cenozoic Glacial Ages* ed K K Turekian (New Haven: Yale University Press) pp 71–181

[2] Guiot J and de Vernal A 2007 Transfer Functions: Methods for Quantitative Paleoceanography

Based on Microfossils *Proxies in Late Cenozoic Paleoceanography (Developments in Marine Geology* vol 1) ed C Hillaire-Marcel and A de Vernal (Amsterdam: Elsevier) chapter 13 pp 523–63

[3] CLIMAP Climate: Long-Range Investigation, Mapping, and Prediction Project Members 1976 The surface of the Ice Age Earth *Science* **191** 1131–37

[4] Gersonde R, Crosta X, Abelmann A and Armand L 2005 Sea-surface temperature and sea ice distribution of the Southern Ocean at the EPILOG Last Glacial Maximum—a circum-Antarctic view based on siliceous microfossil records *Quat. Sci. Rev.* **24** 869–96

[5] ter Braak C J F 1985 Correspondence analysis of incidence and abundance data: properties in terms of a unimodal response model *Biometrics* **41** 859–73

[6] Roux M 1979 Estimation des paléoclimats d'après l'écologie des foraminifères *Cah. Anal. Données* **4** 61–79

[7] ter Braak C J F, Juggins S, Birks H J B and van der Voet H 1993 Weighted averaging partial least squares regression (WA-PLS): definition and comparison with other methods for species-environment calibration *Multivariate Environment Statistics* ed G P Patil and C R Rao (Amsterdam: Elsevier) pp 519–53

[8] Morey A E, Mix A C and Pisias N G 2005 Planktonic foraminiferal assemblages preserved in surface sediments correspond to multiple environmental variables *Quat. Sci. Rev.* **24** 925–50

[9] Hutson W H 1980 The Agulhas current during the late Pleistocene: Analysis of modern faunal analogs *Science* **207** 64–6

[10] Williams J W and Shuman B 2008 Obtaining accurate and precise environmental reconstructions from the modern analog technique and North American surface pollen dataset *Quat. Sci. Rev.* **27** 669–87

[11] Waelbroeck C, Labeyrie L, Duplessy J-C, Guiot J, Labracherie M, Leclaire H and Duprat J 1998 Improving past sea surface temperature estimates based on planktonic fossil faunas. *Paleoceanography* **13** 272–83

[12] Gavin D G, Oswald W W, Wahl E R and Williams J W 2003 A statistical approach to evaluating distance metrics and analog assignments for pollen records *Quat. Res.* 60 356–67

[13] de Vernal A, Turon, J-L and Guiot J 1994 Dinoflagellate cyst distribution in high latitude environments and development of transfer function for the reconstruction of sea-surface temperature, salinity and seasonality *Can. J. Earth Sci.* **31** 48–62

[14] Bartlein P J, Prentice I C and Webb T III 1986 Climatic response surfaces from pollen data for some eastern North American taxa *J. Biogeogr.* **13** 35–57

[15] Racca J M J and Prairie Y T 2004 Apparent and real bias in numerical transfer functions in palaeolimnology J. Paleolimnol. **31** 117–24

[16] Brewer S, Guiot J, Sanchez-Goni M F and Klotz S 2008 The climate in Europe during the Eemian : a multi-method approach using pollen data *Quat. Sci. Rev.* **27** 2303–15

Foraminifera isotopic records... with special attention to high northern latitudes and the impact of sea-ice distillation processes

Claude Hillaire-Marcel

GEOTOP, Université du Québec à Montréal, P.O. Box 8888, succursale "centre ville" Montréal, Qc, H3C 3P8 Canada

E-mail: hillaire-marcel.claude@uqam.ca

Abstract. Since the reassessment of oxygen isotope paleotemperatures by N. Shackleton in the late 60s, most papers using isotopic records from planktic or benthic foraminifers imply a direct relationship between oxygen isotopes in seawater and the ice/ocean volume, thus some linkage with salinity, sea level, etc. Such assumptions are also made when incorporating "isotopic modules" in coupled models. Here, we will further examine the linkages between salinity and oxygen isotope ratios of sea-water recorded by foraminifers, and their potential temporal and spatial variability, especially in the northern North Atlantic and the Arctic oceans. If temporal and spatial changes in the isotopic composition of precipitations and ice meltwaters tune the isotopic properties of the fresh water end-member that dilutes the ocean, rates of sea-ice formation and evaporation at the ocean surface play a further role on the salt and oxygen isotope contents of water masses. Thus, the oxygen 18-salinity relationship carries a specific isotopic signature for any given water mass. At the ocean scale, residence time and mixing of these water masses, as well as the time dependent-achievement of proxy-tracer equilibrium, will also result in variable recordings of mass transfers into the hydrosphere, notable between ice-sheets and ocean. Since these records in water mass may vary in both amplitude and time, direct correlations of isotopic records will potentially be misleading. Implications of such issues on the interpretation of oxygen isotope records from the sub-arctic seas will be discussed, as well as the inherent flaws of such records due to sedimentological and or ecological parameters.

1. Introduction

It is not the intention here to review the literature about stable oxygen and carbon isotopes in the ocean and their recording by biogenic carbonates such as foraminifers. Review papers exist (e.g., [1]). We presume that some knowledge of such proxies that constitute the basics of paleoceanography is now secured in the geoscience community. The present summary will thus address departures from basic principles, with special attention to processes of particular importance in the northern North Atlantic and the Arctic Ocean. Foraminiferal isotopic records have intrinsic limitations linked to foraminifer ecological properties and/or sedimentological processes. Such limitations are of concern everywhere, however high latitude records are also impacted by the effect of sea-ice production rates on the ^{18}O-salinity properties of water masses. In addition, the Atlantic Meridional Overturning Circulation (AMOC) has been strongly impacted by freshwater pulses from the growth and decay of the large continental ice-sheets of the northern hemisphere. Such events have left characteristic imprints in oxygen isotope records that may deserve further examination to be properly interpreted in terms of hydrographical changes in the North Atlantic. These are the few aspects we will examine below in more detail.

2. From oxygen isotopes in calcite to water mass properties

The so-called paleotemperature equation most currently used [2] is as follows:

$$t\,(°C) = 16.9 - 4.38\,(\delta_c - A) + 0.10\,(\delta_c - A)^2$$

with $A = (\delta_w - 0.27)$, δ_c and δ_w standing for the oxygen isotope composition in δ-values against VPDB and VSMOW (see [3]), and the 0.27‰ correction standing for the offset, on the VPDB scale, between the CO_2 equilibrated with VSMOW at 25°C and that extracted from the VPDB, at the same temperature.

The task of the paleoceanographer has been to use this equation with the two unknowns: the temperature and the isotopic composition of the ambient water at the time of carbonate growth, assuming that isotopic equilibrium was indeed achieved during precipitation. Fortunately, many foraminifers currently used in paleoceanographic reconstructions seem to form their calcite under such conditions. Noticeable exceptions exist, such as *Cibicidoïdes wuellerstorfi*. It nonetheless follows other benthic foraminiferal records with a quite constant 0.64‰ offset [4], still not fully explained.

In the near future, we hope to see instrumental capabilities for the analysis of rare isotopic combinations, such as ^{13}C-^{18}O (e.g., mass 47 in the CO_2 extracted from calcite), to be analyzed from very small samples. This would provide a quantitative assessment of the temperature during carbonate precipitation (cf. "clumpted isotope temperatures"; e.g., [5]). Currently, paleotemperatures extracted from elemental ratios such as Mg/Ca from foraminifer shells remain the best way to provide the above equation with a reliable temperature value (see [6] for a review).

Assuming that the growth temperature of the foraminiferal calcite analyzed is known, one then faces the challenge of correctly interpreting the δ_w value calculated from the paleotemperature equation. Since Shackleton's papers in the late 1960s [7], direct linkages between the isotopic composition of seawater and its salinity have been postulated. Indeed, the growth of continental ice sheets represents a distillation process where isotopically light fresh water is removed from the ocean, only to produce opposite trends when returned as melt water. Some departure from this "ideal" budget of mass-transfers within the hydrosphere has been highlighted by Waelbroeck and others [8]. As a matter of fact, the temporal and spatial variability of the isotopic composition of ice meltwater and freshwater,, and the oceanic distribution of water masses with specific properties are likely to result in local isotopic records. These records may show significant departures from any stacked record, whatever the "representativity" of this record may be (e.g., [9, 10]). Site-specific interpretations of such isotopic offsets in terms of changes in salinity are similarly questionable. Distillation processes produced by sea-ice formation at high latitudes, and evaporative processes at low latitudes, result in salt-isotope properties of water masses. These processes produce isotopic values that deviate from any simple linear relationship between local freshwaters and standard mean ocean water. In this overview, we will further examine the processes linked to sea-ice growth.

3. Isotopic imprint of the distillation of sea-water through sea-ice formation

As illustrated below in Figure 1, the linkage between salinity and isotopic composition of water masses from the North Atlantic is more complicated than a simple mixing system between freshwater and seawater.

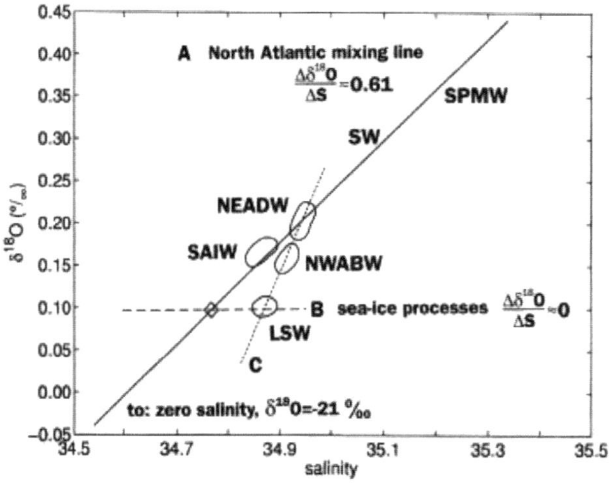

Figure 1. Isotopic composition vs salinity of some major North Atlantic Water masses. SPMW: Sub-Polar Mode Water (~ surface water -SW- of the NW Atlantic); NEADW: North-East Atlantic Deep Water (~Iceland-Faroe-Scotland overflow); NWABW: North-West Atlantic Bottom Water (~Denmark Strait overflow water); LSW: Labrador Sea Water; SAIW: Sub-Arctic Intermediate Water. Note that the surface water line, (A) with its -21‰ y-intercept, does not correspond to a direct mixing line between Standard Mean Ocean Water (SMOW) and continental waters from high latitudes, since the mean isotopic composition of Arctic rivers is of about -19‰. Thus, even this recirculated water mass shows a δ^{18}O-salinity relationship significantly modified by distillation processes resulting in non-conservative properties. Reprinted from [11], with permission from Elsevier.

At low latitudes, deviation from simple mixing lines between SMOW and precipitation waters is caused by kinetic effects during water evaporation, affecting the δ^{18}O-salinity properties of surface waters, such as those of the Gulf Stream. However at high latitudes, distillation of sea-ice from the low salinity, low δ^{18}O surface water layer notably of the Arctic Ocean, results in the production of isotopically light brines as illustrated in figures 2 and 3 below.

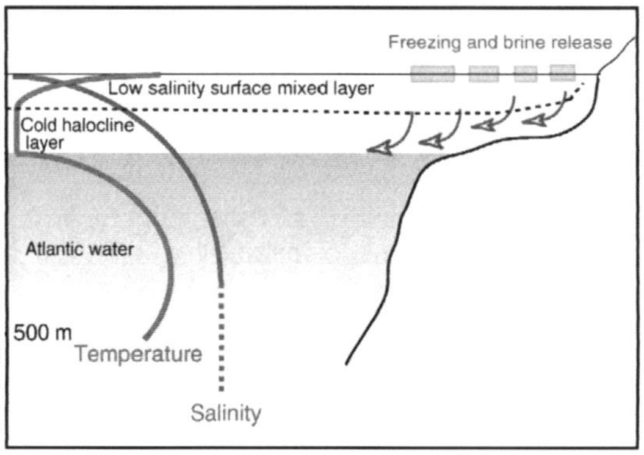

Figure 2. Sketch (modified from [12]) illustrating sea-ice production (with subsequent release of brines sinking in the water column) in the low-salinity (low δ^{18}O) surface water layer of the Arctic Ocean.

The distillation of sea-ice thus releases brines even slightly more depleted in heavy isotopes than the surface water mass itself due to the slight ^{18}O-enrichment of ice vs water which sink in the water column. When melting, sea-ice contributes to the low salinity of the surface water layer while enriching it with heavy isotopes [13, 14]. These processes are illustrated in the figure below.

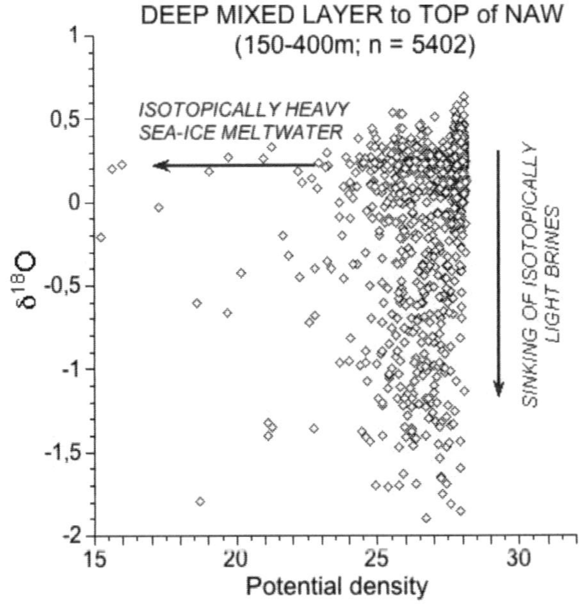

Figure 3. Isotopic composition vs potential density of the subsurface water mass of the Arctic Ocean (on top of the "warm" and saline North Atlantic Water mass, where asymbiotic planktic foraminifers may find a suitable habitat). Data illustrating conditions North of Fram and Bering straits (from [15]).

Neogloboquadrina pachyderma is the only species of planktic foraminifera present in the Arctic. It has also been abundant throughout glacial intervals in North Atlantic records. This planktic species, as well as benthic foraminifers living under areas of sea-ice production, capture these salinity related isotopic offsets. This is illustrated below using isotopic measurements of core-top assemblages of *N. pachyderma* in the Arctic Ocean.

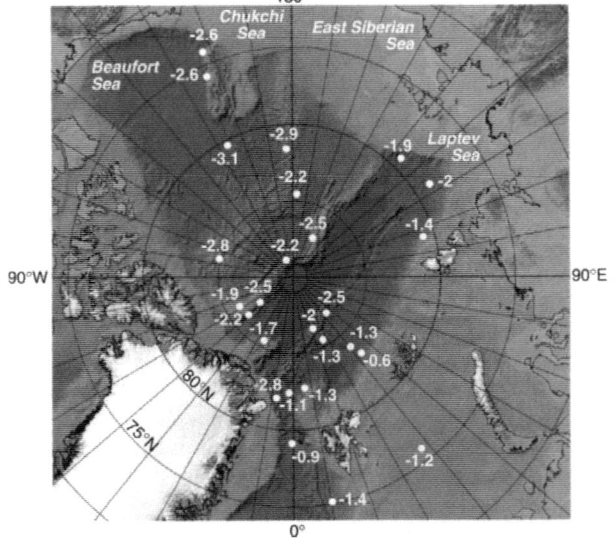

Figure 4. Distribution of isotopic offsets in *N. pachyderma* vs equilibrium conditions for a calcite precipitated at mid-depth along the pycnocline between the cold and dilute surface water layer of the Arctic Ocean and the underlying North Atlantic Water mass. These isotopic offsets have been tentatively linked to the rate of production of isotopically light brines linked to sea ice formation (see references in [16]). The Arctic topography map is from the International Bathymetric Chart of the Arctic Ocean IBCAO).

Based on these observations, one must thus question the forcing process behind the large isotopic excursions recorded by planktic foraminifers in the past, specifically during Heinrich events [16]. Although freshwater pulses such as the final drainage of Lake Agassiz into the NW North Atlantic approximately 8.4 ka ago have practically left no isotopic imprint on planktic foraminiferal records [17, 18], episodes of sea-ice formation that occurred in the North Atlantic during Heinrich events resulted in characteristic "light" isotope shifts (e.g., [18]).

In the example illustrated below, from Jennings et al. [19], several records from the Denmark Strait area depict such large amplitude light δ^{18}O-peaks in both planktic and benthic foraminifers during the Younger Dryas cold spell. *N. pachyderma* in fact reach values as low as 1.6‰. On one hand, the release of huge amounts of fresh water during this interval seems improbable, but on another hand, such a high dilution of the surface water would be incompatible with salinity requirements of *N. pachyderma* (> 34; cf. [20]). Thus, the most likely explanation here is that these fossil assemblages recorded pulses in sea-ice growth with significant addition of ^{18}O-depleted brines to *N. pachyderma* habitat, deeper in the water column or in brine-rich micro-habitat in the ice itself.

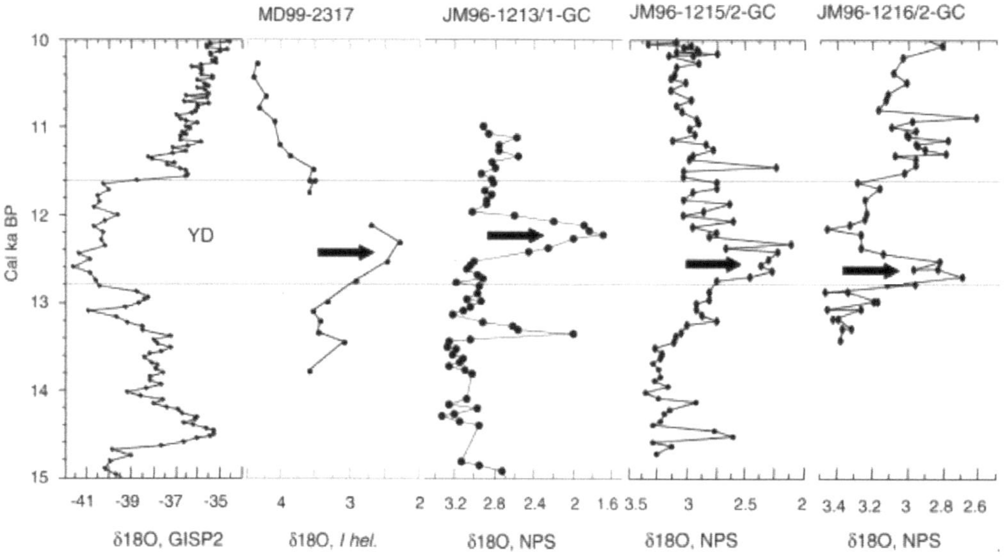

Figure 5. Records spanning the Younger Dryas interval from the Greenland Shelf, east of Denmark Strait, with reference to the GISP2 ice record [21]. Note the large light isotopic excursions (blue arrows), which might indicate episodes of enhanced sea-ice formation. Reprinted from [19], with permission from Elsevier.

A subsequent concern is the dispersal pattern of an isotopic/salinity event in the ocean and thus, the correlation and interpretation of isotopic records in general. Firstly, it has been shown that "major gradients exist, laterally, between the abyssal North Atlantic and North Pacific, and vertically over much of the ocean, persisting for periods longer than 2000 years and with magnitudes bearing little or no relation to radiocarbon ages" [22]. Furthermore, both the glacial and interglacial oceans cannot be seen in an equilibrium state: transient states seem to characterize even relatively long interglacial stages (e.g., [23]). Thus, linkages between calculated δ^{18}Ow-values from foraminiferal records, with specific source-events from glaciers, continents and ocean reorganization cannot be made without care. They might be valid near the sources, but questionable in distal sites.

4. Inherent limitation in foraminiferal and sedimentological records

Several other properties of isotopic and sedimentological records must be kept in mind when using such records as proxy for paleohydrographical conditions, regardless of the latitude of the study site and basin. They are briefly evoked here. A more exhaustive examination of such intrinsic limitations of paleoceanographical records can be found in [1] and [24].

Among major parameters, with respect to records from planktic foraminifers: the seasonality and the interannual/interdecadal variability in their production is particularly important at high latitudes, where they often live near boundaries of their optimal habitat. Similarly, their ecological affinities and depth habitat are quite critical. Most asymbiotic species depict some isopycnal distribution where T and S may vary while preserving the density the living formainifers require. With respect to benthic foraminifers, special attention should also be paid to their specific micro-habitat from sediment surface to a few centimeters below. Then, linked to benthic life in general, the mixing of sediment occurs with a depth-pattern and intensity which are linked to several parameters (organic carbon content, oxygen availability, bathymetry, benthic ecological structure, turbidity, grain size...), thus varying strongly in time and space. This process results in a redistribution of proxy-indicators in the sediment, along with some smoothing and transformation of records, which are further accentuated by variable fluxes of carrier organisms and particles.

5. Moving forward and conclusions

All aforementioned limitations must not be seen as making isotopic records from foraminifer assemblages of little use in paleoceanography. They still remain one of the best proxies for the reconstruction of past-ocean conditions. Thus, if limitations do exist in terms of unequivocal interpretation of isotopic offsets and of their timing, when linked to properties of given water masses or in combination with other tracers, stable isotopes yield coherent scenarios of hydrographical changes at sub-basin or basin scale.

For example, stable isotope data combined with radiogenic and U-series isotope studies have yielded some large features with respect to the Holocene inception of the modern AMOC pattern as illustrated in Figure 6.

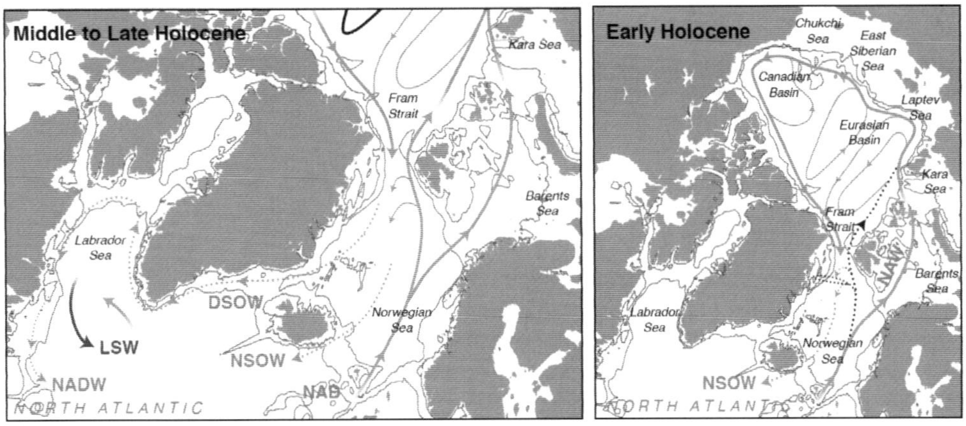

Figure 6. The behavior of the AMOC during the Middle and Late Holocene (left) and the Early Holocene (right). Intermediate or deep water masses are orange; the incoming North Atlantic warm and saline water mass (North Atlantic Drift (NAD) then North Atlantic Water (NAW)) is red. Cold and dilute surface currents evacuating Arctic fresh waters westward are not shown. Note the strong opposition between a high-salinity northeast North Atlantic and a low-salinity northwest North Atlantic. Maximum outflow of Norwegian Sea Overflow Water (NSOW) occurred during the Early Holocene, whereas Denmark Strait Overflow Water (DSOW) peaked

during the mid-Holocene and Labrador Sea Water (LSW) formation reached a maximum during the late Holocene. This east-west temporal shift is linked to increasing density of surface waters westward. Under the condition of increasing freshwater fluxes from the Arctic, the most sensitive sector of deep-intermediate formation would thus be the Labrador Sea, as also indicated by recent modelling experiments. From [25]. Reprinted with permission from AAAS.

For a given water-mass, at a basin scale, it is also possible to derive some information about the evolution of this water mass, as in the example of the Labrador Sea Water illustrated in Figure 7. Due to temperature requirement, *N. pachyderma* occupies the lower pycnocline of the Holocene North Atlantic, on top of the Labrador Sea Water (LSW) since its inception approximately 7 to 8 ka ago (see [26] and references therein). The trends of oxygen isotope compositions from records illustrating several sectors of the North Atlantic converge to a "modern" value of about 2.6‰, suggesting that the complete homogenization of the LSW required a few thousand years, as suggested by Wunsch and Heimbach's experiments [22].

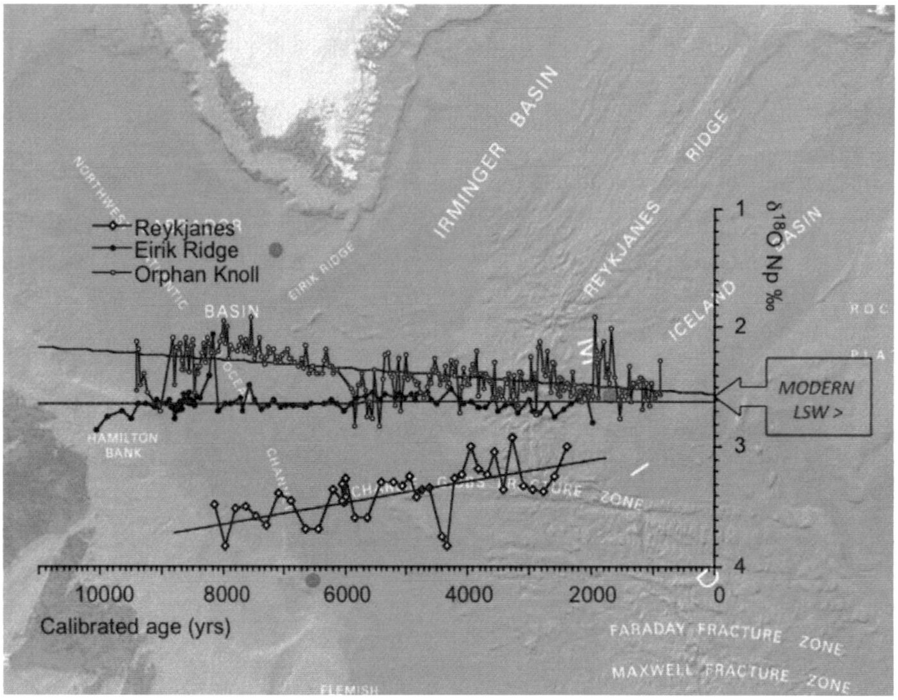

Figure 7. Isotopic composition of *N. pachyderma* in three sites of the North Atlantic suggesting a trend for homogenization of isotopic compositions on top of the LSW mass (*N. pachyderma*'s habitat) (data from [26]). Orphan Knoll: southernmost site.

The examples above provide arguments to continue documenting isotopic compositions in foraminiferal assemblages, especially at basin to sub-basin scales. The caveat, in terms of extrapolation of interpretative lines at a larger scale, is even more critical when looking at nearshore records. In the example of Iceland, the presence and temporal and spatial variability of several water masses around the island may result in "isotopic" ambiguities, where any isotopic signal could indeed respond to some excursion of a nearby water mass or to changes in the properties of a given water mass. In this case, multiproxy approaches will be essential.

Acknowledgments
Many thanks to Dorothea Bauch (GEOMAR, Kiel), and to Antje Volker (Geological Survey, Lisbon) for helpful comments on the first version of this manuscript.

References

[1] Ravelo C and Hillaire-Marcel C 2007 The use of oxygen and carbon isotopes of foraminifera in paleoceanography *Proxies in Late Cenozoic Paleoceanography (Developments in Marine Geology* vol 1*)* ed C Hillaire-Marcel and A de Vernal (Amsterdam: Elsevier) chapter 18 pp 735–64

[2] Shackleton N J 1974 Attainment of isotopic equilibrium between ocean water and the benthonic foraminifera genus *Uvigerina*: isotopic changes in the ocean during the last glacial *Les méthodes quantitatives d'étude des variations du climat au cours du Pléistocène* ed J Labeyrie (Paris: Centre National de la Recherche Scientifique) pp 203–09

[3] Coplen T 1994 Reporting of stable hydrogen, carbon, and oxygen isotopic abundances *Pure Appl. Chem.* **66** 273–76

[4] Shackleton N J and Opdyke N D 1973 Oxygen isotope and palaeomagnetic stratigraphy of Equatorial Pacific core V28-238: Oxygen isotope temperatures and ice volumes on a 10^5 year and 10^6 year scale *Quat. Res.* **3** 39–55

[5] Eiler J M 2007 "Clumped-isotope" geochemistry–The study of naturally-occurring, multiply-substituted isotopologues *Earth Planet. Sci. Lett.* **262** 309–27

[6] Barker S, Cacho I, Benway H and Tachikawa K 2005. Planktonic foraminiferal Mg/Ca as a proxy for past oceanic temperatures: A methodological overview and data compilation for the Last Glacial Maximum *Quat. Sci. Rev.* **24** 821–34

[7] Shackleton N J 1967 Oxygen isotope analyses and Pleistocene temperatures reassessed. *Nature* **215** 15–7

[8] Waelbroeck C, Labeyrie L, Michel E, Duplessy J-C, McManus J F, Lambeck K, Balbon E and Labracherie M 2002 Sea-level and deep water temperature changes derived from benthic foraminifera isotopic records *Quat. Sci. Rev.* **21** 295–305

[9] Lisiecki L E and Raymo M E 2005 A Pliocene-Pleistocene stack of 57 globally distributed benthic $d^{18}O$ records *Paleoceanography* **20** PA1003 doi:10.1029/2004PA001071

[10] Lisiecki L E and Raymo M E 2009 Diachronous benthic $d^{18}O$ responses during late Pleistocene terminations *Paleoceanography* **24** PA3210 doi:10.1029/2009PA001732

[11] Frew R D, Dennis P F, Heywood K J, Meredith M P and Boswell S M 2000 The oxygen isotope composition of water masses in the northern North Atlantic *Deep-Sea Res.* **47** 2265-2286

[12] Aagaard K 1981 On the deep circulation of the Arctic Ocean *Deep-Sea Res.* **28** 251–68

[13] Bédard P, Hillaire-Marcel C and Pagé P 1981 ^{18}O modeling of freshwater inputs in Baffin Bay and Canadian Arctic coastal waters Nature **293** 287–93

[14] Bauch D, Schlosser P and Fairbanks R G 1995 Fresh-water balance and the source of deep and bottom waters in the Arctic Ocean inferred from the distribution of H_2O-^{18}O *Progr. Oceanogr.* **35** 53–80

[15] Schmidt, G. A., Bigg, G. R., Rohling, E. J., 1999. Global seawater oxygen-18 database, http://data.giss.nasa.gov/o18data/

[16] Hillaire-Marcel C and de Vernal A 2008 Stable isotope clue to episodic sea ice formation in the glacial North Atlantic *Earth Planet. Sci. Lett.* **268** 143–50

[17] Hillaire-Marcel C, de Vernal A and Piper D J W 2007 Lake Agassiz Final drainage event in the northwest North Atlantic *Geophys. Res. Lett.* 34 L15601 doi:10.1029/2007GL030396

[18] Hillaire-Marcel C, Hélie J-F, McKay J and de Vernal A 2008 Elusive isotopic properties of deglacial meltwater spikes into the North Atlantic: Example of the final drainage of Lake Agassiz *Can. J. Earth Sci.* **45** 1235–42

[19] Jennings A E, Hald M, Smith M and Andrews J T 2006 Freshwater forcing from the Greenland Ice Sheet during the Younger Dryas: Evidence from southeastern Greenland shelf cores *Quat. Sci. Rev.* **25** 282–98
[20] Hilbrecht H 1997 Morphologic gradation and ecology in *Neogloboquadrina pachyderma* and *N. dutertrei* (planktic foraminifera) from core top sediments *Mar. Micropaleontol.* **31** 31–43
[21] Grootes P M, Stuiver M, White J W C, Johnsen S and Jouzel J 1993 Comparison of oxygen isotope record from the GISP2 and GRIP Greenland ice cores *Nature* **366** 552–54
[22] Wunsch C and Heimbach P 2008 How long to oceanic tracer and proxy equilibrium? *Quat. Sci. Rev.* **27** 637–51
[23] de Vernal A and Hillaire-Marcel C 2008 Natural variability of Greenland climate, vegetation and ice volume during the last million years *Science* **320** 1622–25
[24] Hughen K A 2007 Radiocarbon dating of deep-sea sediments *Proxies in Late Cenozoic Paleoceanography (Developments in Marine Geology* vol 1*)* ed C Hillaire-Marcel and A de Vernal (Amsterdam: Elsevier) chapter 5 pp 193–218
[25] Weaver A J and Hillaire-Marcel C 2004 Global warming and the next ice-age *Science* **304** 400–2
[26] de Vernal A and Hillaire-Marcel C 2006 Provincialism in trends and high frequency changes in the northwest North Atlantic during the Holocene *Global Planet. Change* **54** 263–90

Geochemical proxies of ocean circulation and weathering inputs: Radiogenic isotopes of Nd, Pb, Sr, Hf, and Os

Martin Frank

IFM-GEOMAR, Leibniz Institute of Marine Sciences at the University of Kiel, Wischhofstrasse 1-3, 24148 Kiel, Germany

E-mail: mfrank@ifm-geomar.de

Abstract. Marine records of the radiogenic isotope composition of the elements neodymium (Nd), lead (Pb), hafnium (Hf), strontium (Sr), and osmium (Os) allow the reconstruction of past continental weathering inputs on different time scales as a function of their respective oceanic residence times. Sr and Os have oceanic residence times significantly longer than the global mixing time of the ocean and are efficiently mixed on a global scale. Their isotope composition changes on long time scales as a function of plate tectonics and major orogenies, which allows their use as precise stratigraphic tools for the entire Phanerozoic. In contrast, Hf, Pb, and in particular Nd, have residence times on the order of or shorter than the global mixing time of the ocean, which results in distinct isotopic signatures of water masses and allows the reconstruction of past water mass mixing and weathering inputs on both long and short time scales. Here applications of these isotopes systems with a focus on the shorter residence time tracers are reviewed (without claiming to be comprehensive) and problems and potential solutions are discussed. **Keywords: Radiogenic isotopes, paleo-oceanography, ocean circulation, water mass mixing, continental weathering**

1. Introduction

The radiogenic isotope signatures in rocks vary because one of the isotopes has changed its abundance over time due ingrowth caused by decay of a radioactive parent isotope (i.e. ^{143}Nd is the decay product of ^{147}Sm, which means that the ratio between ^{143}Nd and primordial ^{144}Nd is variable). Consequently the ^{143}Nd/^{144}Nd ratio of the rocks changes as a function of their age and the ratio between the concentrations of samarium (Sm) and Nd. The radiogenic isotope composition of dissolved trace metals in the ocean primarily reflects the balance between the radiogenic isotope composition of the continental weathering inputs and mantle derived sources, either weathering of mantle-derived basaltic rocks or hydrothermal contributions (Figure 1) (see [1,2] for a review).

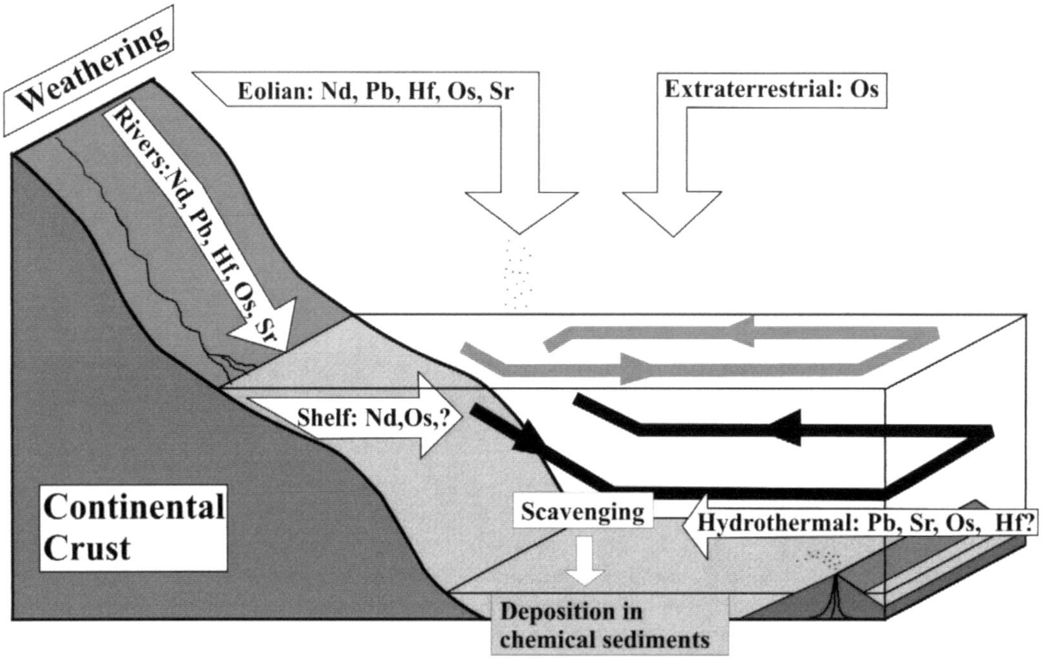

Figure 1. Sources, sinks and pathways of trace metals with radiogenic isotope signatures.

Strontium and osmium have long oceanic residence times between about 10,000 years and 2 million years, respectively. Their radiogenic isotope compositions are consequently homogenously distributed in global seawater. Stored in authigenic carbonates, such as microfossil shells Sr and Os isotope records provide information about long term changes in weathering inputs as a function of tectonic changes, such as major orogenies or changes in the production rate of oceanic crust as a consequence of the break-up of continents. Due to these systematic global variations, which have been recorded at many locations and on many geological time scales, Sr isotopes have successfully been used for establishing a reliable marine stratigraphy over the entire Phanerozoic [3]. The global data base for variations in the Os isotope composition of global seawater is currently much smaller (e.g. [4]) but the data base is constantly growing [5] and Os isotopes have for example successfully been used for dating marine hydrogenetic ferromanganese crusts [6], in which Sr isotope variations are not reliably preserved (e.g. [7,8]).

In contrast, Nd, Hf and Pb are particle reactive in seawater and thus their concentrations and radiogenic isotope compositions are not homogenously mixed in global seawater. These elements have residence times between 50-200 years (Pb, [9]) and 400 to 2000 years (Nd, Hf, [10,11,12]) which is on the order of or shorter than the global ocean mixing time. In the open ocean, where these metals are not directly influenced by weathering inputs, the radiogenic isotope systems of Nd and Hf can serve as isotopic fingerprints of water masses and thus quasi-conservative tracers of water mass mixing [12,13,14,15,16,17]. The water masses acquire their isotopic signatures in the source areas of the water masses through weathering inputs and subsequently only change their isotope composition by mixing with other water masses if circulation is vigorous enough, as is the case for Nd isotopes in the present day Atlantic Basin (Figure 2). These radiogenic isotope signatures are not influenced by any biological fractionation, which makes them valuable paleo proxies of past ocean circulation and weathering inputs provided that suitable archives are available.

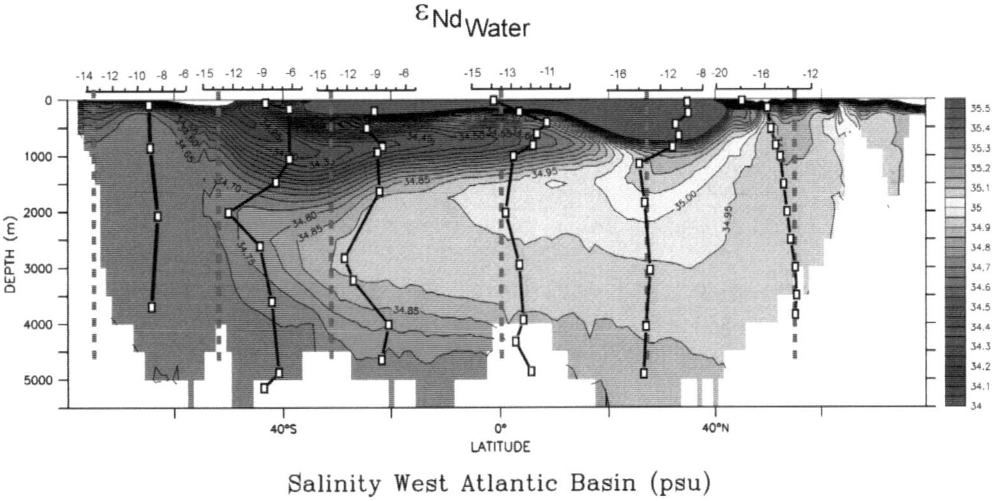

Figure 2. Water mass distribution on a N-S transect in the present day Atlantic Ocean as characterized by salinity (in ‰) [18]. North Atlantic Deep Water (NADW) flowing south is characterized by its high salinity and is compensated by northward flowing cold Antarctic Bottom Water (AABW) and warmer low salinity Antarctic Intermediate Water (AAIW). Water column ε_{Nd} data are superimposed, which clearly reflect this distribution of the water masses [14,15,16] (ε_{Nd} values represent the deviation of the measured $^{143}Nd/^{144}Nd$ of a sample from that of the Chondritic Uniform Reservoir (CHUR) in parts per 10,000). The figure has been reprinted with permission from [16]. Copyright 1999 American Association for the Advancement of Science

The radiogenic isotope composition of past seawater is either recorded in chemical sediments or seawater-derived chemical components, which incorporate the radiogenic isotope composition of seawater during their growth. Suitable chemical sediments recording long term changes in the radiogenic isotope composition of past deep waters are hydrogenetic ferromanganese crusts or nodules (see [1] for a review). Records from these chemical precipitates are generally restricted to a relatively coarse time resolution due to their extremely low growth rates on the order of few millimeters per million years. Advances in resolution of these archives were achieved through laser ablation measurements of Pb and Nd isotopes [19,20], which, however, still only allow a resolution on the order of 10,000 years per measurement making it difficult to reliably reconstruct glacial-interglacial variations.

More recent efforts have therefore been focussed on the extraction of the radiogenic isotope signatures of seawater from archives permitting higher time resolution records. It has been shown for Nd isotopes that foraminiferal shells deposited in marine pelagic sediments can serve as archives for deep water [21] and even surface water Nd isotope signatures of the past [22,23,24,25]. The paleo surface water records have for example been used to reconstruct changes in riverine inputs or erosional signals in the Labrador Sea or the Bay of Bengal on millennial to million year time scales. The extraction of the Nd isotope signatures, in particular those of surface waters, is, however, complicated by the fact that the concentrations within the foraminiferal carbonate itself are extremely low. Despite the fact that several recent studies have suggested that the methods to acquire surface water data of the past are now reliable [e.g. 26] the method still needs further studies to determine the effects of potential diagenetic additions of Nd from deep waters with a different isotope signature [27].

Other archives that have been successfully and reliably used to reconstruct the past Nd isotope composition of deep waters are fossil fish teeth, which incorporate the deep water Nd isotope composition during early diagenesis of the phosphate at the sediment water interface [28,29,30], and more recently deep water corals, which incorporate the Nd into their carbonate skeleton [31,32,33,34]. In the past ten years the most commonly used archive to extract deep water radiogenic isotope compositions for reconstructions of past weathering inputs and ocean circulation has been thin ferromanganese coatings on foraminifera or other sediment particles, which form through early diagenetic processes at the sediment water interface [35,36,37]. At particular locations the reliability of this archive is also currently under debate due to effects of sediment redistribution or partial dissolution of volcanic glasses during leaching of the sediments.

2. Applications and case studies: Progress and problems

2.1. Pb and Hf isotopes: Tracing past weathering regimes and continental inputs

Pb has a very short average residence time in seawater due to its high particle reactivity and consequently cannot be transported very far without vertical additions. The isotopic composition of Pb shows a large range of variations in continental rocks and is thus a suitable tracer for nearby continental weathering inputs in the past and with some restrictions also for short range water mass exchange in the past [38,39]. While the Nd isotope composition of a source rock is essentially transferred into seawater without large changes, the Pb and Hf isotope compositions are fractionated during weathering on land (incongruent weathering) and therefore also carry information about the prevailing weathering regime near the source regions of particular water masses. In the case of Pb, the controlling process has been the preferential release of highly radiogenic Pb, which is weakly bound in the mineral lattices due to radiation damage, from rock surfaces freshly eroded by glaciers [40]. Based on this process, Pb isotope time series from sediments in the western North Atlantic have for example been used to reconstruct the past retreat of the Laurentide ice sheet since the Last Glacial Maximum and thus the transition from dominantly glaciated landscapes with reduced chemical weathering to the modern weathering regime [41].

The fractionation of Hf isotopes, in contrast, predominantly occurs because weathering-resistant zircons contain large amounts of unradiogenic Hf, which are not released during chemical weathering, but can only significantly be introduced into seawater when glacial grinding of rocks releases a significant fraction of the Hf contained in the zircons (see [42] for a review). This process, together with preferential release of radiogenic Hf from some labile minerals [43,44] and potential hydrothermal inputs of Hf [45] are mainly responsible for the fact that seawater and its archives for a given Nd isotope signature show a more radiogenic Hf isotope composition if compared to expected values from bulk rock analyses. In addition, the weathering effects result in a reduction of the global range of Hf isotope ratios in seawater compared with bulk rock signatures. A trend towards more congruent weathering in the North Atlantic Ocean over the past 2-3 million years has been revealed by combined Hf-Nd isotope analyses of ferromanganese crusts from the western North Atlantic, which has been interpreted as a consequence of weathering conditions increasingly controlled by physical weathering during glacial periods since the major intensification of Northern Hemisphere Glaciation at 2.7 million years ago [46].

2.2. Nd isotopes: Reconstructing past ocean circulation

The main interest in radiogenic isotope systems for paleo-oceanography, however, arises from the fact that radiogenic isotope systems can be used to constrain past water mass mixing on ocean basin scales. The most suitable system for this purpose is that of Nd, which can be measured precisely both in seawater and sedimentary archives and can provide clear information about the water mass structure, such as for example in the Atlantic Ocean (Figure 2).

As a case study the evolution of the export of NADW from the North Atlantic to the Southern Ocean, one of the key currents of the global thermohaline circulation system, is discussed here. Such reconstructions have been carried out on different time scales. On the million years time scale ferromanganese crust records in the North Atlantic have revealed a substantial shift of the Nd isotope signature of NADW to less radiogenic values over the past 3 million years [38,47], probably mainly caused by increased supply of unradiogenic Nd from Northern Canada and Greenland in the course of the major intensification of Northern Hemisphere Glaciation [47]. Over the same period of time this shift was not mirrored by similar changes of the Southern Ocean's main water mass, the Circumpolar Deep Water (CDW) [48], into which NADW is ultimately entrained and which today constitutes about 50-70 % of the Nd in CDW, the other 30-50 % being of Pacific origin with much higher Nd isotope signatures [49]. This absent shift was interpreted as an overall decrease of NADW export by 15-35% over the past 3 million years although the observed Nd isotope variations of the Southern Ocean crusts may entirely have been caused by a progressively stronger decrease of NADW export only during glacial periods because samples from ferromanganese crusts integrate over several 10,000 years of growth history [48].

The major interest in past changes of the export of NADW is on much shorter millennial to centennial time scales. The first higher resolution record of CDW for the past 70,000 years was obtained from the ferromanganese coatings of a sediment core in the Cape Basin, which clearly revealed much more Pacific-like Nd isotope signatures of CDW during glacial Marine Isotope Stages (MIS) 4 and 2 [50]. In agreement with previously obtained benthic carbon isotope records this was interpreted as a significant decrease in NADW export to the Southern Ocean during glacial periods. Further work on the same sediment core at even higher, up to centennial scale resolution revealed similarly marked decreases in NADW export during the cold episodes (stadials) of MIS 3 [51]. While these stadial/interstadial excursions in Nd isotopes occurred synchronously with the benthic carbon isotope changes measured on the same samples, thus excluding uncertainties arising from the chronostratigraphy of the core, a phase lag has been observed at the main MIS transitions, in particular when going from warm to cold stages [51]. The records indicate that benthic oxygen isotope signatures and thus global ice volume started to change first, followed by benthic carbon isotopes some 2,000-3,000 years later, which were affected by changes in the carbon cycle. Lastly, Nd isotopes changed, again about 2,000 years later, indicating that changes of the ocean circulation cannot have triggered the sequence of events at these major climatic transitions but were only a consequence of other forcing factors [51].

A major prerequisite for the validity of these conclusions is that the Nd isotope composition and concentration of Nd of the water masses from the Pacific and of NADW contributing to CDW have remained constant over at least the past 100,000 years. In the case of the Pacific mixing endmember, this is hard to prove by data due to a lack of suitable archives but given the size of the Pacific Basin and its dominant weathering inputs from mantle-derived rocks the assumption of a near constant endmember composition for this period of time seems justified. For the NADW, archives are available to reconstruct its Nd isotope composition at high time resolution. A laser ablation Nd isotope record of two ferromanganese crusts that grew in the water depth of present day NADW was just sufficient to resolve glacial and interglacial Nd isotope signatures of NADW and did not indicate major changes over the past 500,000 years [20]. A similar conclusion was obtained from Nd isotope data of deep sea corals collected within present-day NADW in the western North Atlantic [31], although it was not possible to find corals that grew during the last glacial maximum. A third set of records was also generated from ferromanganese coatings of sediments from Blake Nose on the western North Atlantic continental slope [52]. These drift sediments deposited under vigorous current speeds were strongly affected by sediment redistribution (focusing). Sediment particles deposited in water depths of highest along slope velocities showed least influence by sediment focusing and closely reflected present day Nd isotope signatures of NADW at a nearby location [15], whereas the shallower last glacial main boundary current indicated Nd isotope values of $\varepsilon Nd = -10$. This either reflects a significantly less radiogenic signature of the glacial equivalent of NADW, the Glacial North Atlantic Intermediate Water (GNAIW) [52], or a periodically stronger influence of intermediate waters from the Southern Ocean [53], or a combination of both. The reason for these discrepant results on the past Nd isotope composition of NADW is not yet clear but the reconstruction of past water masses from continental slope sediments may at least partly be influenced by exchange of the bottom waters with the slope sediments themselves, a phenomenon described as boundary exchange [54].

At the same time, glacial deep waters in the western Atlantic have become significantly less radiogenic in their Nd isotope composition due to enhanced northward flow of deep water masses of Southern Ocean origin [52,55]. The study of [55] also revealed that leaching of bulk sediments may at some locations in the North Atlantic result in too radiogenic Nd isotope compositions and that leaching of only the carbonate fraction may lead to more reliable results depending on location.

In summary, it is obvious that more groundtruthing studies are required in this rapidly emerging field of chemical oceanography/paleoceanography. This in particular, is the case for the methods applied to extract the radiogenic isotope signature of past water mass signatures from marine sediments, which need to be developed further in order to be able to exploit their full potential in paleo-ceanography. While at many locations, even with complex detrital sediment compositions, bulk sediment leaching worked well and provided reliable results [see above,56,57], this is not the case for others. In addition, sediment redistribution and downslope transport may bias a reliable reconstruction of water masses for example from sediments on continental slopes [52,58]. There is, however, also a clear need for a significant increase in the global data base for the present day dissolved and particulate Nd and Hf isotope distribution in the ocean, both in the frame of process studies at the margins of the oceans, but maybe more importantly along basin wide ocean sections. This is required to better understand the relationships of these isotope systems to present day water mass distribution and ocean circulation and is currently being investigated by the international GEOTRACES program [59].

References

[1] Frank M 2002 *Rev. Geophys.* **40** 1001, 10.1029/2000RG000094
[2] Goldstein SL and Hemming SR 2003 Long-lived isotopic tracers in oceanography, paleoceanography, and ice-sheet dynamics *Treatise on Geochemistry* ed H Elderfield (Oxford: Elsevier) pp 453–489
[3] Veizer J, Ala D, Azmy K, Bruckschen P, Buhl D, Bruhn F, Carden GAF, Diener A, Ebneth S, Godderis Y, Jasper T, Korte G, Pawellek F, Podlaha OG and Strauss H 1999 *Chem. Geol.* **161** 59
[4] Peucker-Ehrenbrink B and Ravizza G 2000 *Terra Nova* **12** 205
[5] Klemm V, Frank M, Levasseur S, Halliday AN and Hein JR 2008 *Earth Planet. Sci. Lett.* **273** 175
[6] Klemm V, Levasseur S, Frank M, Hein JR and Halliday AN 2005 *Earth Planet. Sci. Lett.* **238** 42
[7] Ling H-F, Burton KW, O'Nions RK, Kamber BS, von Blanckenburg F, Gibb AJ and Hein JR 1997 *Earth Planet. Sci. Lett.* **146**, 1
[8] Henderson G and Burton KW 1999 *Earth Planet. Sci. Lett.* **170** 169
[9] Schaule BK and Patterson CC 1981 *Earth Planet. Sci. Lett.* **54** 97
[10] Tachikawa K, Jeandel C and Roy-Barman M 1997 *Earth Planet. Sci. Lett.* **170** 433
[11] Arsouze T, Dutay J-C, Lacan F and Jeandel C 2009 *Biogeosciences* **6** 2829
[12] Rickli J, Frank M and Halliday AN 2009 *Earth Planet. Sci. Lett.* **280** 118
[13] Zimmermann B, Porcelli D, Frank M, Rickli J, Lee D-C and Halliday AN 2009 *Geochim. Cosmochim. Acta* **73** 91
[14] Jeandel C 1993 *Earth Planet. Sci. Lett.* **117** 581
[15] Piepgras DJ and Wasserburg GJ 1987 *Geochim. Cosmochim. Acta* **51** 1257
[16] von Blanckenburg F 1999 *Science* **286** 1862
[17] Porcelli D, Andersson PS, Baskaran M, Frank M, Björk G and Semiletov I 2009 *Geochim. Cosmochim. Acta* **73** 2645
[18] Levitus S 1982 *NOAA Prof. Pap.* **13** 1
[19] Christensen JN, Halliday AN, Godfrey LV, Hein JR and Rea DK 1997 *Science* **277** 913
[20] Foster GL, Vance D and Prytulak J 2007 *Geology* **35**, 37
[21] Klevenz V, Vance D and Mezger K 2008 *Earth Planet. Sci. Lett.* **265** 571
[22] Vance D and Burton KW 1999 *Earth Planet. Sci. Lett.* **173** 365
[23] Burton KW and Vance D 2000 *Earth Planet Sci. Lett.* **176** 425
[24] Vance D, Scrivner AE, Beney P, Staubwasser M Henderson GM and Slowey NC 2004 *Paleoceanography* **19** PA2009, 10.1029/2003PA000957

[25] Osborne AH, Vance D, Rohling EJ, Barton N, Rogerson M and Fello N 2008 *Proceedings of the National Academy of Sciences of the United States of America* **105** 16444
[26] Martínez-Botí MA, Vance D and Mortyn PG 2009 *Geochem. Geophys. Geosyst.* **10** Q08018, 10.1029/2009GC002701
[27] Pomiès C, Davies GR, Conan SMH 2002 *Earth Planet. Sci. Lett.* **203** 1031
[28] Martin EE and Haley BA 2000 *Geochim. Cosmochim. Acta* **64** 835
[29] Scher HD and Martin EE 2006 *Science* **312** 428
[30] Martin E, Blair SW, Kamenov GD, Scher HD, Bourbon E, Basak C and Newkirk DN 2010 *Chem. Geol.* **269** 414
[31] van de Flierdt T, Robinson LF, Adkins JF, Hemming SR and Goldstein SL 2006 *Paleoceanography* **21** PA4102, 10.1029/2006PA001294
[32] van de Flierdt T, Robinson LF, Adkins JF 2010 *Geochim. Cosmochim. Acta* **74** 6014
[33] Copard K, Colin C, Douville E, Freiwald A, Gudmundsson G, De Mol B and Frank N 2010 *Quaternary Science Reviews* **29** 2499
[34] Colin C, Frank N, Copard K and Douville E 2010 *Quaternary Science Reviews* **29** 2509
[35] Rutberg RL, Hemming SR and Goldstein SL 2000 *Nature* **405** 935-938
[36] Bayon G, German CR, Boella RM, Milton JA, Taylor RN, Nesbitt RW 2002 *Chem. Geol.* **187** 170
[37] Gutjahr M, Frank M, Stirling CH, Klemm V, van de Flierdt T and Halliday AN 2007 *Chem. Geol.* **242** 351
[38] Burton KW, Ling H-F and O'Nions RK 1997 *Nature* **386** 382
[39] Abouchami W, Galer SJG and Koschinsky A 1999 *Geochim. Cosmochim. Acta* **63** 1489
[40] von Blanckenburg F and Nägler TF 2001 *Paleoceanography* **16** 424
[41] Gutjahr M, Frank M, Halliday AN and Keigwin LD 2009 *Earth Planet. Sci. Lett.* **286** 546
[42] van de Flierdt T, Goldstein SL, Hemming SR, Roy M, Frank M and Halliday AN 2007 *Earth Planet. Sci. Lett.* **259** 432
[43] Bayon G, Vigier N, Burton KW, Carignan J, Etoubleau J and Chu N-C 2006 *Geology* **34** 433
[44] Godfrey LV, Zimmermann B, Lee D-C, King RL, Vervoort JD, Sherrell RM and Halliday AN 2009 *Geochem. Geophys. Geosyst.* **10** Q08015, 10.1029/2009GC002508
[45] Bau M and Koschinsky A 2006 *Earth Planet. Sci. Lett.* **241** 952
[46] van de Flierdt T, Frank M, Lee DC and Halliday AN 2002 *Earth Planet. Sci. Lett.* **198** 167
[47] O'Nions RK, Frank M, von Blanckenburg F and Ling H-F 1998 *Earth Planet. Sci. Lett.* 155 15
[48] Frank M, Whiteley N, Kasten S, Hein JR and O'Nions RK 2002 *Paleoceanography* **17** 1022, 10.1029/2000PA000606
[49] Piepgras DJ and Wasserburg GJ 1982 *Science* **217** 207
[50] Rutberg RL, Hemming SR and Goldstein SL 2000 *Nature* **405** 935
[51] Piotrowski AM, Goldstein SL, Hemming SR and Fairbanks RG 2005 *Science* **307** 1933
[52] Gutjahr M, Frank M, Stirling CH, Keigwin LD and Halliday AN 2008 *Earth Planet. Sci. Lett.* **266** 61
[53] Pahnke K, Goldstein SL and Hemming SR 2008 *Nature Geoscience* **1** 870
[54] Lacan F and Jeandel C 2005 *Earth Planet. Sc. Lett.* **232** 245
[55] Roberts NL, Piotrowski AM, McManus JF and Keigwin LD 2009 *Science* **327** 75
[56] Haley BA, Frank M, Spielhagen RF and Eisenhauer A 2008 *Nature Geoscience* **1** 68
[57] Gourlan AT, Meynadier L, Allègre CJ, Tapponnier P, Birck J-L and Joron J-L 2010 *Quaternary Science Reviews* **29** 2484
[58] Stumpf R, Frank M, Schönfeld J and Haley BA 2010 *Quaternary Science Reviews* **29** 2462
[59] SCOR working group 2007 *Chemie der Erde – Geochemistry* **67** 85

Application of Visible/near Infrared derivative spectroscopy to Arctic paleoceanography

Joseph D. Ortiz

Kent State University, Department of Geology, Kent, OH 44242

E-mail: jortiz@kent.edu

Abstract. The lack of well-preserved carbonate in much of the Arctic marine environment dictates the need for alternative methods of paleoceanographic reconstruction. The broad variety of physical properties measurements makes them well suited for use in a variety of environments, but they provide unique opportunities when employed in the Arctic. Because Arctic sediment is introduced and reworked by a variety of mechanisms, the signature from multiple processes becomes intermixed with the sediment. Many of these processes operate in other ocean basins, while some function only in Polar Regions. A strategy to address this mixing problem is to employ spectrally-resolved physical properties measurements, or to use multiple methods in conjunction to generate multivariate data sets, which can differentiate concurrent processes. Data of this type is well suited to multivariate analysis techniques such as sample-based or variable-based, varimax-rotated, principle component analysis (VPCA). These are methods that decompose the data matrix to infer process from orthogonal functions. The method is applied to cores from the Chukchi sea to document that visible derivative spectroscopy provides a powerful means of reconstructing sediment provenance. In the Chukchi Sea, diffuse spectral reflectance provides a proxy to monitor variations in Holocene flow through the Bering Strait.

1. Introduction

Many paleoceanographic studies rely on measurements extracted from the coarse component (>63 μm) of carbonate-bearing, marine sediment. Faunal analysis of planktic and benthic foraminfera constrain past changes in temperature, productivity, and community structure. Benthic foraminiferal oxygen isotopes provide stratigraphic information useful for correlation between cores, as well as insights into the magnitude of terrestrial ice sheets, the temperature of surface and deep water masses, and when used in conjunction with concurrent temperature estimates, proxy information on salinity. Accelerator Mass Spectrometry (AMS) ^{14}C analysis of biogenic carbonate provides information about water mass ventilation age or stratigraphic age control over the past 50,000 cal B.P., while a host of geochemical tracers yield insights into temperature, nutrient content, and ocean circulation. But, what can be done in environments that lack carbonate? Often cores from these "*mare incognita*" environments - where terrigenous input is high - or carbonate preservation is poor - were left unanalyzed because age models were wanting and researchers lacked the necessary tools to generate intelligible proxies.

In settings of this type, often found in shallow coastal environments, deep water, Antarctic or Arctic deposits, use of sediment physical properties provides an effective means of extracting paleoceanographic information. The usefulness of physical properties measurements has become increasingly realized as new tools such as the GEOTEK Multisensor core logger were introduced, the sample size needed for AMS ^{14}C measurements has decreased, and new dating approaches such as the paleomagnetic intensity curve were developed. Because this topic is so broad, this paper will focus on visible near infrared derivative spectroscopy (VNIR) as one example of how physical properties can be used to extract quantitative, lithologic information from marine sediment, but the reader should keep in mind that similarly useful information can be extracted from other physical properties of marine sediment. For more in depth discussion of these methods, the reader is referred to [1, 2].

2. Regional setting and mechanisms of Arctic sediment transport

The Arctic is essentially a land locked ocean basin. For this reason, it is sometimes referred to as the Arctic Mediterranean Sea [3]. Seawater exchanges between the Arctic and other ocean basins via Fram Strait, the Bering Strait, and the Canadian Archipelago [4]. Significant freshwater input and export from the Arctic occurs in the form of precipitation, riverine flux, sea ice and iceberg discharge [5,6]. Sediment transport mechanisms common in other parts of the world (e.g. eolian, fluvial input, down slope sediment transport by mass wasting, and turbidity currents) are also common in the Arctic [7,8,9]. Additional high latitude mechanisms of sediment transport active in the Arctic include nepheloid flows resulting from sinking brines that entrain sediment during sea ice formation and transport of ice-rafted debris (IRD) by glacial icebergs and marine sea ice [10,11,12]. Sediment-laden, sea ice forms through either "anchor ice formation" in shallow water where sea ice can freeze to the bottom, uprooting poorly sorted sediment, or suspension freezing in which fine-grained sediment in fresh water plumes is entrained during sea ice formation [13]. The sediment preserved on the Arctic seafloor thus holds a rich story of sediment provenance and climate change, if appropriate proxies are employed to unmix and extract this information.

3. Physical properties measurements

Sediment physical properties reflect the lithology of the formation and the characteristics of the associated pore fluids. Researchers have a number of variables from which to select depending on the question of interest, instrumentation available and temporal resolution required. These include parameters such as sediment reflectance in the visible and near infrared, thermal conductivity, sonic velocity, density, natural gamma decay, and magnetic properties. Resistivity measurements provide insights into the properties of the pore fluids such as whether they are polar or non-polar, and their level of mobility. While useful physical properties studies can be completed using visual observation, modern instrumentation provides quantitative results with superior reproducibility and the ability to rapidly generate large data sets with exceptional accuracy and precision. Studies can be conducted on the formation using seismic or well logging, on sediment cores using hand held or automated core logging tools, or on laboratory-extracted samples. Useful physical properties can be measured by mechanical or acoustic means, electrical response, and nuclear or magnetic properties. Instruments can either passively detect a spontaneous formation property, or actively induce a response using an energetic source and an appropriate detector. Non-invasive measurements have the advantage of speed and minimal sample preparation, but the disadvantage that the instrument response is often controlled by more than one variable, which complicates interpretation and potentially leads to non-unique solutions. Multivariate statistical methods described below provide a means of overcoming this limitation. It is important to match the instrumental method to the question in terms of variable and resolution. Because we often cannot measure exactly the property of interest, it is best to use multiple tools, or spectrally resolving instruments in conjunction with multivariate statistics to help constrain results.

4. Putting the Electromagnetic spectrum to work

Radiation in the electromagnetic (EM) spectrum can be characterized by its energy, wavelength, or frequency, the reciprocal of wavelength. Low energy components of the EM spectrum, such as radio waves, have long wavelength and low frequency, while high energy gamma rays have short wavelength and high frequency. High energy x-rays and gamma rays have considerable penetration power making them ideal to map out internal composition and sediment fabric, or to determine variations in bulk density.

Energy in the visible and near infrared (VNIR) part of the EM spectrum lies between these two extremes. The absorption pattern of VNIR radiation varies depending on sediment mineralogy and is thus useful for determining lithology [14, 15, 16, 17, 18, 19]. To use data sets of this type to their maximum requires some theoretical knowledge about the nature of VNIR radiation, and multivariate methods, which is described in this section.

In the visible part of the EM spectrum, radiation interacts with matter through a variety of electronic effects, which correlate across a wide range of wavelengths. This is an important characteristic that can be exploited by several types of multivariate statistical methods to decompose a visible reflectance spectrum into independent components. In the NIR, vibration and rotation are the two potential mechanisms by which radiation is absorbed [20]. However, because the mineral crystal lattice tends to inhibits rotation, the most important means by which sediment absorbs NIR is generally vibration. Unlike visible radiation, NIR vibrations are band-limited to a narrow wavelength range due to the nature of the bonds between elements within the crystal lattice. This means that the correlation between NIR bands that are far apart tends to be very low. This difference has important implications for analysis of visible and NIR data by multivariate statistical methods as described below.

As discussed above, multivariate data often exhibits intercorrelation between variables. It is generally not possible to build reliable statistical models from intercorrelated variables because the model may try to explain the same variability more than once by inserting correlated variables into the relationship. This problem, called overfitting, arises from collinearlity or multicolinearity within the data set. These terms are used for bivariate and multivariate data respectively [21]. Multivariate data can be analyzed to explore the relationships between variables, or samples depending on the types of data used and the questions of interest.

A useful approach to multivariate analysis includes eigenvalue –eigenvector methods, such as principle component Analysis (PCA) and factor analysis. These approaches are extremely powerful, but their relationship has been poorly understood by many researchers, in part because the methods were developed simultaneously in different fields with distinct jargon. PCA seeks to decompose a data matrix into independent (or orthogonal) components by finding the eigenvalues and eigenvectors of the matrix. This is identical to solving a system of linear equations. Because PCA amounts to a series of matrix operations, it can be applied to extract independent variance between variables, or to extract independent information between samples depending on how the matrix is oriented and normalized. Variable based PCA is similar to R-Mode factor Analysis, while sample based PCA is similar to Q-Mode factor Analysis. Aside from how the matrix is decomposed, the data matrix can be pre-processed (transposed, normalized, standardized, or transformed) and the extracted solution can be post-processed or rotated to improve the power of the model. Reference [21] discussed the relationship between variable based and sample based PCA and factor analysis in further detail for interested readers.

A variety of pre-processing and post-processing steps can be incorporated into PCA. Here we employ a specific type of PCA, Varimax-rotated, principle component analysis (VPCA, which seeks to explain the structure of the variance between variables. Pre-processing steps include deciding whether to transform the data and whether to normalize by subtracting the variable mean or standardize by subtracting the variable mean and divide by its standard deviation. If normalized data is used, it is then transformed to yield the variance-covariance matrix, while the standardized data is transformed to yield the correlation matrix. It is one of these two matrices which is then decomposed by finding its eigenvectors and eigenvalues. The eigenvectors, or principle components, represent linear combinations of the original variables that are independent or orthogonal to each other. The component loadings within each principle component show the relative importance of each original variable in each principle component. The eigenvalues define how much variance is explained by each principle component. The communality explains how much of the variance in each of the original variables each principle component explains. The principle components can be multiplied (or "projected") back onto the original data matrix to produce component scores that explain the relative importance of each component on each sample.

A common post-processing step is to rotate the component solution to aid in the interpretation of the results. This can be done using oblique rotations, which introduce correlation between the principle components, or with an orthogonal rotation which maintains the independence of the components. While oblique rotations are often employed, this method should be avoided at all costs because it violates the fundamental advantage of the method, which is to extract independent components from the inter-correlated data, and thus address the problem of multicolinearity. The use of an oblique rotation reintroduces multicolinearity into the results. The most common orthogonal rotation is the varimax-rotation, which rotates the components around the multivariate central mass or centroid of the data to maximize the importance of the smallest number of variables on the structure of the extracted principle components [22].

Varimax-rotated, PCA is well suited for analysis of visible reflectance data because VPCA seeks to preserve variance within the variables (bands or wavelengths) in a spectrum when the correlation structure of the data is decomposed to yield orthogonal mineral assemblages [22, 23, 24, 25, 26]. In contrast, because NIR measurements are band-limited, there is often low correlation between wavelengths that are not closely spaced. In this case, two samples with the same minerals present will exhibit the same correlated bands, but because NIR features are not present at all bands and the intercorrelation of the bands in an NIR spectrum is lower than in a visible spectrum, variable based, VPCA tends to blow up noise when used with VNIR data, particularly when decomposing a correlation matrix.

A better approach when analyzing VNIR data is thus to employ sample based PCA which seeks to preserve information across variables within a given sample. Unlike variable based PCA modes, sample based PCA modes group like samples, which in this case, share common VNIR features. The resulting sample based PCA has fewer assemblages with less spectral features than a typical variable based PCA model, making it easier to interpret, at the expense of compositional resolution. The reason that sample based PCA returns fewer modes than variable based PCA is because there are almost always many fewer variables than samples in the sample based matrix, than in the variable based matrix that is decomposed. As a result, the sample based method is over-determined; there is much more redundant information in the sample based analysis than in the variable based analysis.

5. VNIR derivative spectroscopy

While there are many effective means through which physical properties can be used to characterize sediment variability, VNIR derivative spectroscopy has several desirable aspects that make the method extremely versatile. Reflectance spectra are sensitive to grain size and moisture content, which can alter the shape of the reflectance spectra by scattering and absorption. These effects can be minimized by calculating the derivative of the reflectance spectrum to remove these matrix effects. Reflectance derivative spectroscopy is diagnostic for carbonates, iron oxides/oxyhydroxides, clay minerals and plant pigments [27, 28, 29, 30, 31, 32]. It relies on lower energy radiation and is thus safer to employ than high energy physical properties methods. Sample analysis can be conducted rapidly with minimal sample preparation. The instrumentation is small enough to be portable, relatively inexpensive, and easy to maintain.

The color of an object depends on its inherent properties, the spectral properties of the illuminating light source, the response of the detector, and the color measurement system used to represent the data. Light that interacts with an object can be scattered, transmitted, or absorbed. The relative proportion of these responses depends on the angle of incidence of the light, the opacity of the material and the configuration of its electrons. Analysis of color is referred to as colorimetry. Colorimetry is fairly broadly used in sedimentology to generate color indices such as lightness (L*), the yellow-blue contrast (b*), and the red-green (a*) contrast, which mimic human color vision and integrate information across the visible part of the spectrum. These indices are reproducible within a given sediment matrix and can be empirically calibrated to measure sediment composition, but color alone is not diagnostic of mineralogy because various mixtures can obtain any specific color. In contrast the reflectance derivative spectrum for a given mineral is diagnostic if the illumination and detector response are controlled. Because the elements in different minerals have varying orbital configurations and bonds, they exhibit different reflectance derivative spectra in the visible and near infrared. Specific features and spectral shapes can be used to identify minerals by comparison of reflectance derivative spectra from an unknown sample with a reference derivative spectrum for various known standards. The method is analogous to quantitative XRD identification [33, 34, 35]. In contrast, reflectance derivative spectroscopy is widespread in remote sensing applications [e.g., 36, 37], but relatively uncommon in marine sedimentology. In the following sections, I present two case studies to document how VNIR derivative spectroscopy can be used to evaluate sediment provenance and climate change.

6. The Chukchi Sea

The complete VPCA analysis consists of a matrix of principle component loadings, which in this case can be plotted against wavelength. These represent scaled assemblages of mineral reflectance derivative spectra. They can be compared against reference mineral reflectance derivative spectra for identification in much the way that one compares XRD measurements from samples with standards. The vector of eigenvalues explains how much information each principle component pattern explains, while the communality vector describes the fraction of variance in each band explained by the sums of squares of the component loadings. The component scores, the projection of the component loadings back onto the matrix of reflectance derivative spectra can be plotted versus depth and provides down core patterns that demonstrate the variability of each principle component through time. We'll apply this approach to samples from the western Arctic to evaluate changes in clay minerals on the Alaskan Margin during the late Quaternary.

The dominant clay mineral in Arctic marine sediment is illite, while the dominant clay mineral in the Pacific is chlorite [38]. In the Western Arctic, water flowing through Bering Strait carries suspended sediment from the Pacific into the Arctic, mixing these two clay minerals [19]. Because these minerals have distinct reflectance spectra, they can be differentiated by VNIR spectroscopy. To monitor changes in the flow through Bering Strait into the Chukchi Sea, seven piston cores were collected on the Alaskan shelf and slope near Barrow Canyon during cruise HLY0501 aboard the USCGC Healy (Figure 1). These samples were augmented with grab and core samples from earlier cruises aboard the USCGC Healy (HLY0403 and HLY0205) to enhance spatial and temporal resolution. Post-cruise measurements of diffuse spectral reflectance (DSR) by visible spectroscopy using a Minolta CM-2600d spectrophotometer were conducted at 1 cm resolution on the lightly scraped surfaces of split cores covered in plastic wrap to protect the instrument's integration sphere. The instrument measures DSR in the visible at 10 nm resolution from 400-700 nm on a 3 mm spot size. The 30 channels of data from nine cores yield 270 down core time series with 12,505 spectra or 375,150 reflectance values. A multivariate set of this size must be simplified to extract useful information. The traditional means of analyzing spectrophotometric data is by colorimetry. The spectral reflectance values are converted to the CIE L*a*b* color indices which reduces the 30 channels of data down to three indices calibrated against human color vision and which can be relate to the brightness (L*), the red-green contrast (a*), and the yellow-blue contrast (b*) of the sediment (Figure 2). The disadvantages of this approach is that, as with human color vision, there is significant correlation between the a* and b* values and the input reflectance variables themselves are also significantly correlated. Because electronic effects, which absorb light in the visible, are not as band limited as vibration effects in the NIR, there is strong inter-correlation of the variance between visible bands, resulting in redundant variability (Figure 3). As a result, there is redundant information included in the L*a*b* color space which makes it impossible to extract theoretical relationships between these indices and mineral composition. They must be empirically calibrated on a site by site basis.

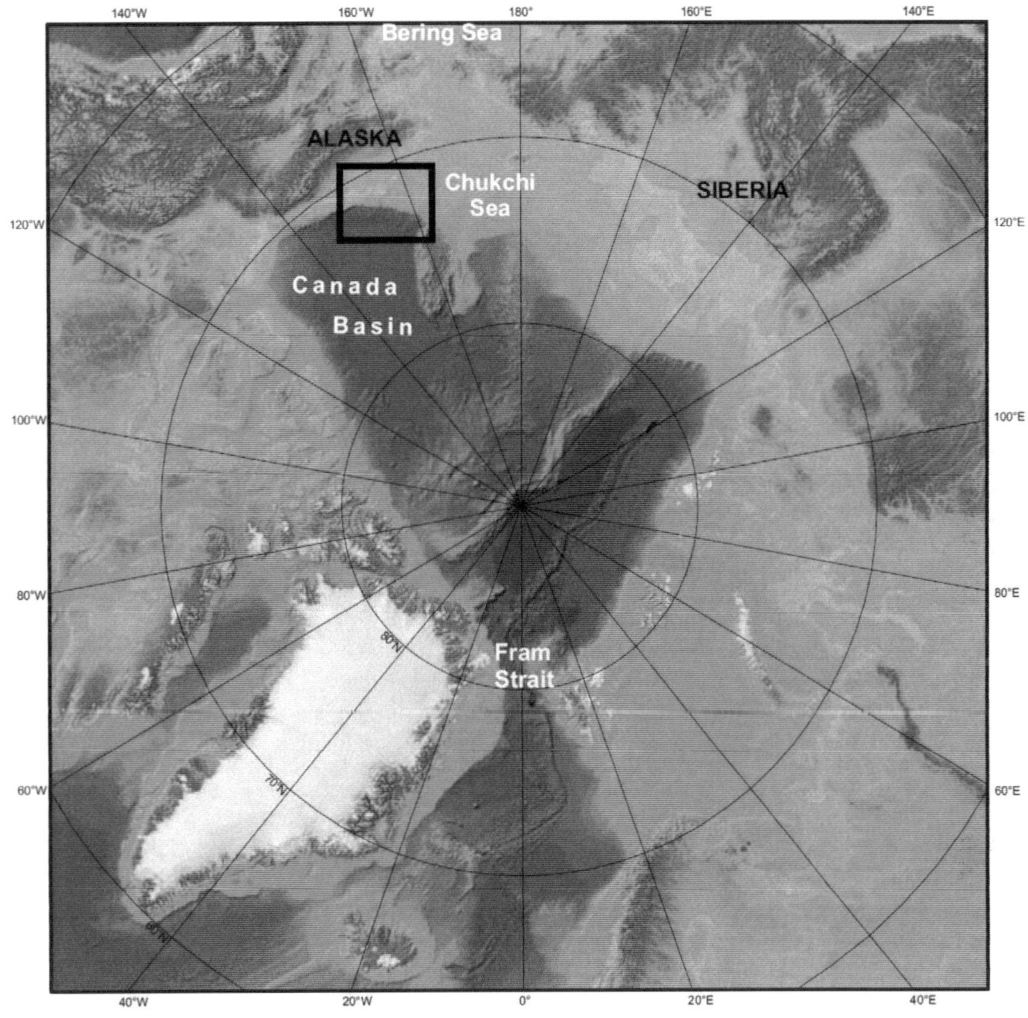

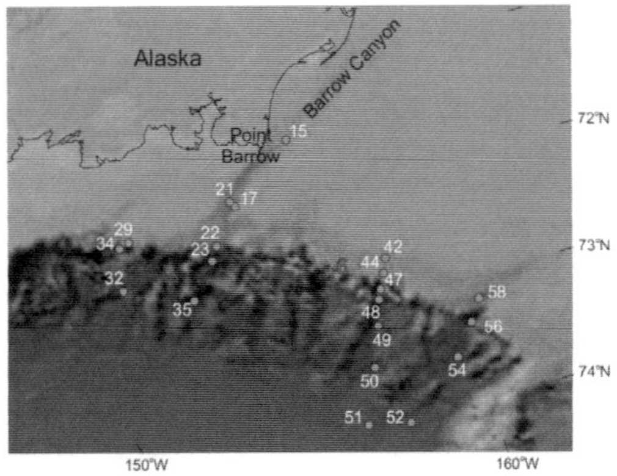

Figure 1. (Top) The IBACO bathymetry [39] of the Arctic Ocean showing the study region in the Chukchi Sea; (left) inset map showing the location for core HLY0501-JPC6 (star), and the HLY0403 shelf basin interaction (SBI) surface samples (circles).

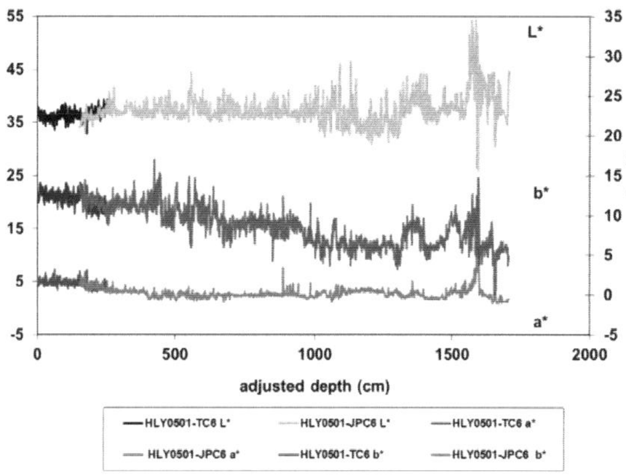

Figure 2. CIE L*A*B* colorimetric indices for the spliced record from cores JPC6 and TC6 plotted against depth using a 147 cm offset for core JPC6 relative to TC6. The upper curve is L*, the middle curve is b* and the lower curve is a*.

To avoid this problem we conduct a PCA analysis of the derivative-transformed reflectance spectra. The derivative transformation removes scattering effects and decreases some of the resulting intercorrelation of the reflectance data (Figure 3). The principal component analysis completes the decomposition. Analysis of the Chukchi sea reflectance data set in this way yields three components verified by comparison with reflectance derivative spectra for mixtures of known mineral standards and by quantitative XRD of selected sediment samples. The three leading modes represent (1) Smectite and Dolomite, (2) Illite and Goethite, and (3) Chlorite and Muscovite (Figure 4). The shape of the reflectance derivative spectra varies in samples with different proportions of these components (Figure 5) The down core variations of these mineral components inferred by derivative spectroscopy

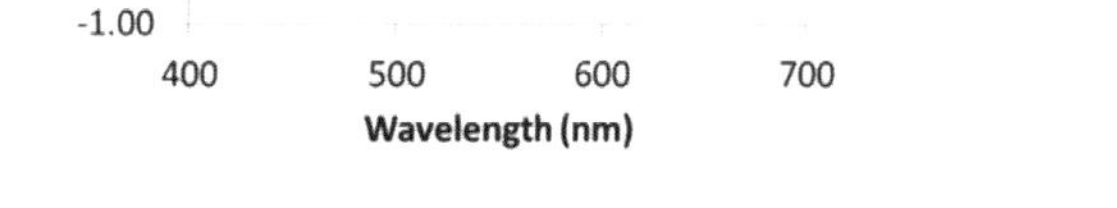

Figure 3. Variation of Pearson's correlation coefficient for 400 nm reflectance (blue) versus other reflectance curves measured at 10 nm resolution from 400 to 700 nm using a Minolta CM-2600d spectrophotometer. The Pearson's correlation the derivative of the reflectance data is plotted in red. Measurements are from core HYL0501-6JPC.

is confirmed by analysis of samples by quantitative XRD (Figure 6). The distribution of these components can be mapped geographically (Figure 7) and as a function of depth (figure 8). These components as expressed off the Alaskan Margin are associated with fine grained minerals transported in suspension, by nepheloid flows, and to some extent, by ice rafted debris [13]. The illite and goethite component is found in greatest abundance on the shelf (Figure 7, 8), potentially delivered to the region as rock flour created by glaciers and reworked by coastal processes. The smectite and dolomite component is found in greatest abundance on the slope (Figures 7, 8), while the chlorite and muscovite component is found at depths just below the shelf break (Figure 7,8). The thickness of the Holocene sediment drape decreases linearly with depth along the shelf, except for cores raised from ~500-600 m water depth along the steepest portion of the slope (Figure 8), and at 1200 m water depth on the slope

(Figure 8). Material transported down the slope to this drift deposit site via Barrow Canyon has the characteristics of the continental shelf, and is enriched in illite and goethite. The Chlorite and Muscovite assemblage gradually increases in abundance in a core raised from the Chukchi slope from

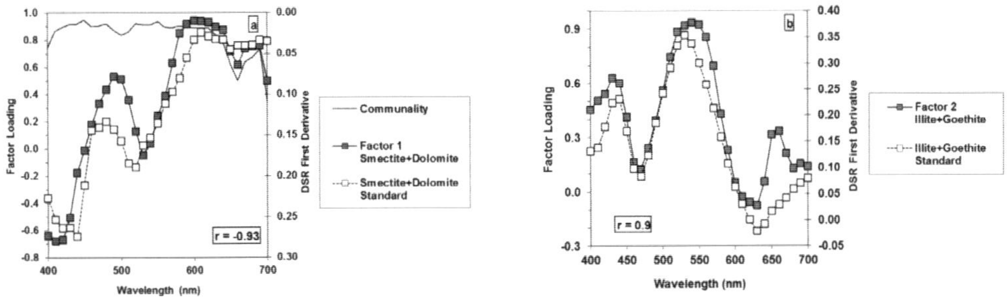

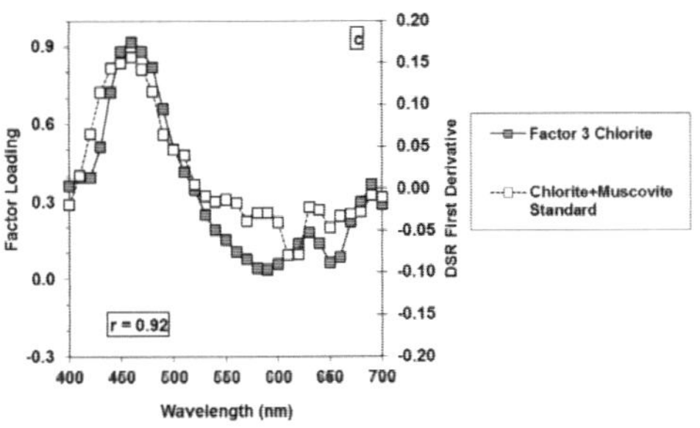

Figure 4. Varimax-rotated component loadings and DSR first derivatives of mineral standards plotted as a function of wavelength. The composition of the Alaskan margin sediment was determined by linear correlation of the factor loadings against reflectance signatures for known samples. (a) Factor 1 loadings vs. smectite+dolomite; (b) Factor 2 loadings vs. illite + goethite (c) Factor 3 loadings vs. chlorite + muscovite.

12,000 years BP until 6,000 years BP when it increases abruptly and then fluctuates until decreasing abruptly 4,000 years ago. This fluctuation corresponds to a re-organization of the global climate system during the mid-Holocene (Figure 9), including a reduction in the strength of the Atlantic Meridional Overturing Circulation [19].

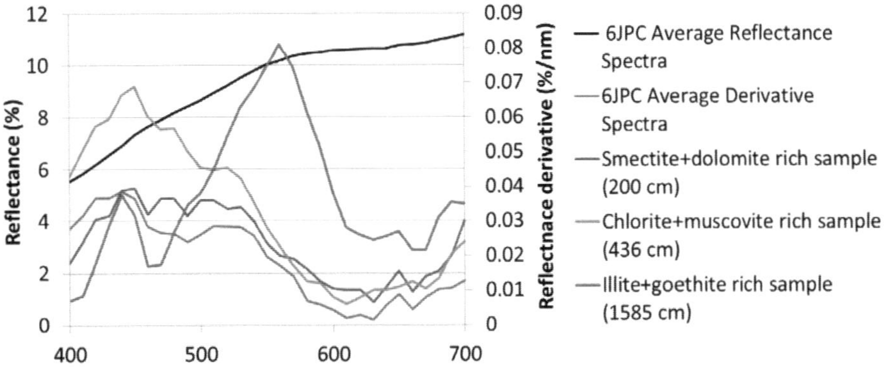

Figure 5. Average reflectance spectrum (black line) and average derivate spectrum (red line) for core HLY0501-6JPC. Selected derivate spectra for samples from various depth correspond to samples rich in smectite+dolomite (blue line), chlorite+ muscovite (green line), and illite+goethite (red line).

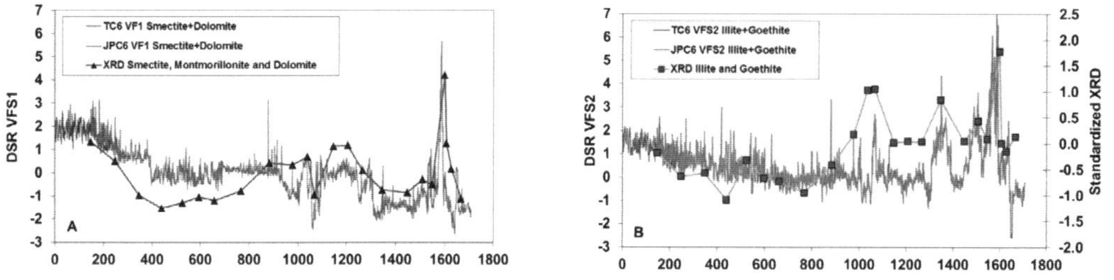

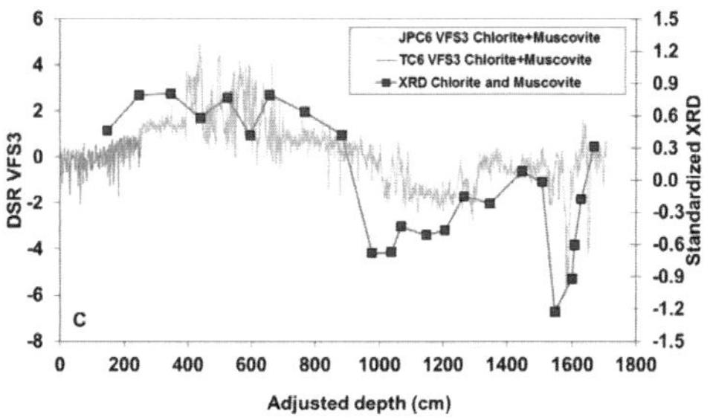

Figure 6. XRD analysis of selected down core samples (filled symbols) from JPC6 confirms the interpretation of the DSR factors (gray lines) presented in Figure 4 and 5. (a) smectite+dolomite, (b) illite+goethite, (c) chlorite+muscovite.

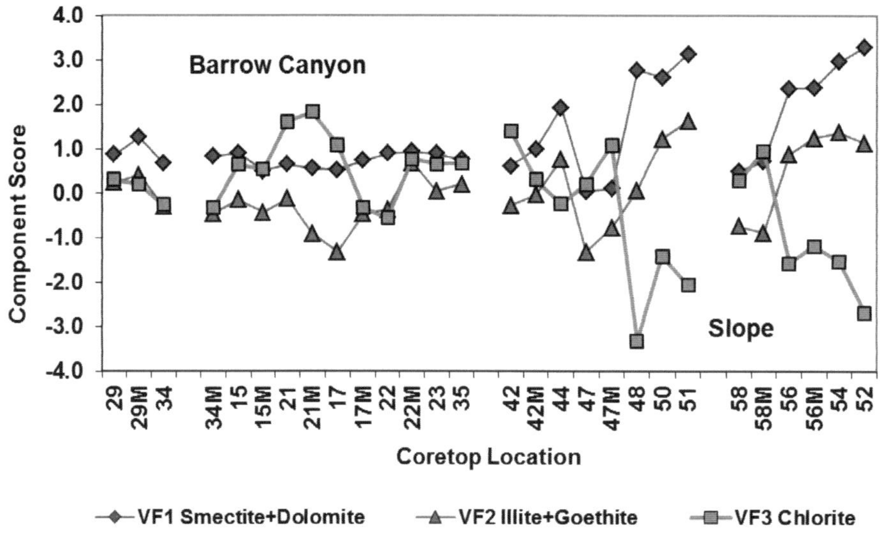

Figure 7. The distribution of varimax factor scores in the HLY0403 core tops aids in the interpretation of the down core patterns for each factor. Transects in the east have moderate illite+goethite and smectite+dolomite. Samples in the western transects have higher illite+goethite and smectite+dolomite. The chlorite+muscovite factor is dominant on the shelf, particularly in Barrow Canyon and decreases in importance down slope.

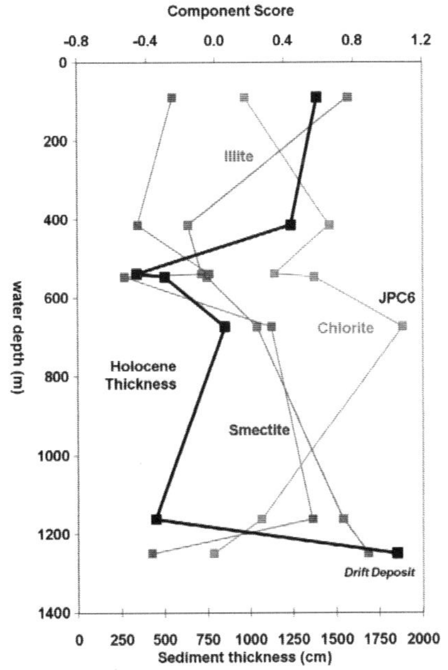

Figure 8. The distribution of the reflectance components plotted as a function of depth on the Alaskan Margin. Sediment thickness is plotted in black, the smectite+dolomite component is plotted in blue. The illite+goethite component is plotted in red, and the chlorite+muscovite component is plotted in green.

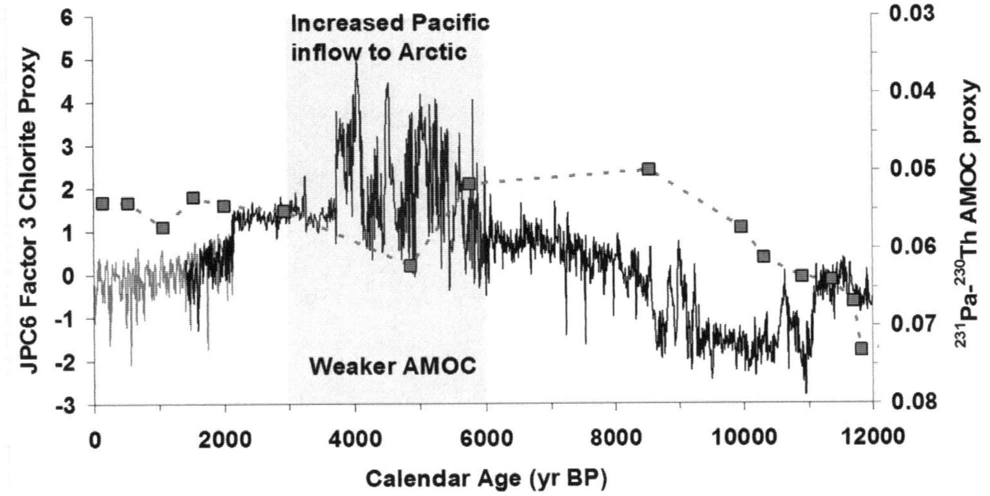

Figure 9. The mid- to late Holocene Varimax component 3 chlorite+muscovite record for spliced core TC6/JPC6 plotted against age. The curve documents enhanced chlorite+muscovite on the Chukchi Shelf between 6000 and 2000 calendar year BP during the mid-Holocene, a time of reorganization within the climate system and a weakening of the AMOC [40].

7. Conclusions

Sediment physical properties can provide a useful means of extracting information from the sedimentary record. The broad variety of physical properties measurements makes them well suited to use in a variety of environments, however, they are particularly important in carbonate-poor Arctic sediment where many traditional paleoceanographic proxies are not as effective. Because physical properties measurements often are sensitive to multiple processes, they are most effective when they provide spectrally resolved data, or when they are used in conjunction with each other. Data of this type is well suited to multivariate analysis techniques such as VPCA. Application of VPCA to diffuse spectral reflectance data of sediment from cores raised from the Alaskan Margin of the Chukchi Sea, documents that derivative spectroscopy provides a powerful means of reconstructing sediment provenance. In the Chukchi Sea, diffuse spectral reflectance provides a proxy to monitor variations in Holocene flow through the Bering Strait and allows for mapping of this variability geographically and temporally.

Acknowledgements

I thank Guillaume St. Onge and Anne de Vernal for inviting me to participate in the HU 2008-029 expedition, and Guillaume St. Onge for the invitation to participate in the 2010 IODP summer school. This research was funded in part by grants from the National Science Foundation Office of Polar Programs (ARC-0612384) and by a grant from the Kent State University Research Council to Ortiz. Chelsea Orsburn and Renee Crane assisted with sample preparation. We thank Lloyd Keigwin for providing us with access to the cores collected during HLY0205 cruise aboard the *USCGC Healy*.

References

[1] Ortiz, J D and Rack F R 1999 Non-invasive sediment monitoring methods: Current and future tools for high-resolution climate studies *Reconstructing Ocean History: A window into the future* ed F Abrantes and A Mix (New York: Kluwer) pp 343–80

[2] St-Onge G Mulder T Francus P and Long B 2007 Continuous physical properties of cored marine sediments. *Proxies in Late Cenozoic Paleoceanography (Developments in Marine Geology* vol 1) ed C Hillaire-Marcel and A de Vernal (Amsterdam: Elsevier) chapter 2 pp 63-98

[3] Dietrich G Kalle K Krauss W and Siedler G 1980 *General Oceanography* 2nd ed. (New York: Wiley Interscience)

[4] Aagaard K Swift J and Carmack E 1985 Thermohaline circulation in the Arctic mediterranean seas, *J. Geophys. Res.* **90** 4833–46

[5] Schoster F Behrends M Muller C and R Stein 2000 Modern river discharge and pathways of supplied material in the Eurasian Arctic ocean: evidence from mineral assemblages and major and minor element distribution *Int. J. Earth Sci.* **89** 486–95

[6] Stein R (Editor) 2008 *Arctic Ocean Sediments: Processes, Proxies, and Paleoenvironment. (Developments in Marine Geology* vol 2) (Amsterdam: Elsevier) 592 pp.

[7] Darby DA Naidu AS Mowatt TC and Jones G 1989 Sediment composition and sedimentary processes in the Arctic Ocean *The Arctic Seas - Climatology, Oceanography, Geology, and Biology* ed Y Herman (New York: Van Nostrand Reinhold) pp 657–720

[8] Nürnberg D Wollenburg I Dethleff D Eichen H Kassens H Letzig T Reimnitz E and Thiede J 1994 Sediments in Arctic sea ice: Implications for entrainment, transport, and release *Mar. Geol.*, **119** 185–214

[9] Wahsner M Müller C Stein R Ivanov G Levitan M Shelekhova E and Tarasov G 1999 Clay-mineral distribution in surface sediments of the Eurasian Arctic Ocean and continental margin as indicator for source areas and transport pathways – a synthesis *Boreas* **28** 215–33

[10] Backhaus J O Fohrmann H Kampf J and Rubino A 1997 Formation and export of water masses produced in Arctic shelf polynyas – process studies of ocean convection *ICES J. Mar. Sci.* **54** 366–82

[11] Detleff D Rachold V Tintelnot M and Antonow M 2000 Sea-Ice transport of riverine particles from the Laptev Sea to Fram Strait based on clay mineral studies *Int. J. Earth Sci.* **89** 496–502
[12] Phillip R L and Grantz A 2001 Regional variations in provenance and abundance of ice-rafted clasts in Arctic Ocean sediments: implications for the configuration of late Quaternary oceanic and atmospheric circulation in the Arctic *Mar. Geo.* **172** 91–115
[13] Darby D A Ortiz J D Polyak L Lund S Jakobsson M and Woodgate R A 2009 The role of currents and sea ice in both slowly deposited central Arctic and rapidly deposited Chukchi-Alaskan margin sediments *Global Planet. Change* **68** 58–72
[14] Balsam W L and Deaton B C 1991 Sediment Dispersal in the Atlantic Ocean: Evaluation by visible light spectra *Rev. Aquatic Sci.* **4** 411–47
[15] Deaton B C and Balsam W L 1991 Visible spectroscopy: a rapid method for determining hematite and goethite concentrations in geological materials *J. Sediment. Petrol.* **61** 628–32
[16] Mix AC Harris S E Janecek T R 1995 Estimating lithology from nonintrusive reflectance spectra: Leg 138 *Proc. of the Ocean Drill. Prog., Scientific Results* vol 138 ed N G Pisias L A Mayer T R Janecek A Palmer-Julson and T H van Andel (College Station, TX: Ocean Drilling Program) pp 413–27
[17] Mix A C Lund D C Pisias N G Boden P Bornmalm L Lyle M and Pike J 1999 Rapid Climate Oscillations in the Northeast Pacific during the last deglaciation reflect northern and southern hemisphere sources *AGU Geophy. Monog.* **112** 127–48
[18] Ortiz J D Mix A Harris S and O'Connell S 1999 Diffuse spectral reflectance as a proxy for percent carbonate content in North Atlantic sediments *Paleoceanography* **14** 171–86
[19] Ortiz J D Polyak L Grebmeier J M Darby DA Eberl D D Naidu S and Nof D 2009 Provenance of Holocene sediment on the Chukchi-Alaskan margin based on combined diffuse spectral reflectance and quantitative X-Ray Diffraction analysis *Global Planet. Change* **68** 73–84
[20] Clark D L Whitman R R Morgan K A and Mackay S D 1980 Stratigraphy and glacio-marine sediments of the Amerasian Basin, central Arctic Ocean *Geol. Soc. Amer. Spec. Paper* **181** (Boulder: The Geol. Soc. of Amer.) 57 pp
[21] Davis J C 2008 *Statistics and Data Analysis in Geology* 3rd ed J (New York: Wiley and Sons) 656 pp
[22] Kaiser H F 1958 The varimax criteria for analytic rotation in factor analysis *Psychometrica* **23** 187–200
[23] Kachigan S K 1991 *Multivariate statistical analysis: a conceptual introduction* (New York: Radius Press) 303 pp
[24] Smith L I 2002 A Tutorial on Principal Components Analysis http://kybele.psych.cornell.edu/;edelman/Psych-465-Spring-2003/PCA-tutorial.pdf (Accessed: June 1, 2010)
[25] Schlens J 2005 A tutorial on Principal Component Analysis http://www.snl.salk.edu/~shlens/pca.pdf (Accessed June 1, 2010)
[26] Broersen A van Liere R and Heeren R M A 2005 Comparing three PCA based methods for the 3D visualization of imaging spectroscopy data *Visualization, Imaging, and Image Processing* ed J J Villanueva (Benidorm, Spain: ACTA Press)
[27] Clark R N Swayze G A Wise R Livo K E Hoefen T M Kokaly R F and Sutley S J 2003 USGS Digital Spectral Library splib05a *USGS Open File Report 03-395*
[28] Jarrard R D and Vanden Berg M D 2006 Sediment mineralogy based on visible and near infrared reflectance spectroscopy *New Techniques in Sediment Core Analysis (Geol Soc. London Special Pub.* vol 267*)* ed R G Rothwell 129–40
[29] Kutser T Pierson D C Tranvik L Reinart A Sobek S and Kallio K 2005 Using Satellite remote sensing to estimate colored dissolved organic matter absorption coefficient in lakes *Ecosystems* **8** 709–20 doi 10.1007/s10021-003-0148-6

[30] Montero Sanchez I C and Brimhall G H 2001 Semi-automated mineral identification algorithm for ultraviolet, visible, and near-infrared reflectance spectroscopy *Proc. 2001 Annual Conference of the International Association for Mathematical Geology* (Kansas: IAMG) June 2001

[31] Viscara Rossel R A McGlynn R N and McBratney A B 2006 Determining the composition of mineral-organic mixes using UV-VIS-NIR diffuse reflectance spectroscopy *Geoderma* **137** 70–82

[32] Yurco L Ortiz J D Polyak L and Darby D 2010 Reflectance based clay mineral cycles in Quaternary sediments from the Northwind Ridge: implications for glacial-interglacial sedimentation patterns in the Arctic Ocean *Polar Res.* **29** 176–97 doi: 10.1111/j.1751-8369.2010.00160.x

[33] Eberl D D 2003 User guide to RockJock-A program for determining quantitative mineralogy from X-ray diffraction data *USGS Open File Report* **OF 03-78** 40 pp

[34] Eberl DD 2004 Quantitative mineralogy of the Yukon River system: variations with reach and season, and determining sediment provenance *Amer. Mineralogist* **89** 1784–94

[35] Will G 2006 *Power Diffraction; the Rietveld Method and the Two-Stage Method* (Berlin: Springer) 224 pp

[36] Martin S 2004 *An Introduction to Ocean Remote Sensing* (Cambridge: Cambridge Univ. Press) 426 pp

[37] McClain C R 2009 A decade of satellite ocean color observations *Annual Rev. Mar. Sci.* **1** 19–42 doi: 10.1146/annurev.marine.010908.13650.

[38] Kalinenko V V 2001 Clay Minerals in Sediments of the Arctic Seas *Lith. and Min. Res.* **36** 362–72 [Translated from *Litologiya I Poleznye Iskopaemye* **4** 418–29]

[39] Jakobsson M Macnab R Mayer M Anderson R Edwards M Hatzky J Schenke H-W and Johnson P 2008 An improved bathymetric portrayal of the Arctic Ocean: Implications for ocean modeling and geological, geophysical and oceanographic analyses *Geophy. Res. Lett.* **35** L07602 doi:10.1029/2008GL033520

[40] McManus J F Francois R Gherardi J-M Keigwin L D and Brown-Leger S 2004 Collapse and rapid resumption of Atlantic meridional circulation linked to deglacial climate changes *Nature* **428** 834–37

AUTHOR INDEX